assessing
student
understanding
in science
second edition

SANDRA K. ENGER   ROBERT E. YAGER

assessing student understanding in science

second edition

A STANDARDS-BASED K–12 HANDBOOK

CORWIN

A SAGE Company

For information:

Corwin
A SAGE Company
2455 Teller Road
Thousand Oaks, California 91320
(800) 233-9936
Fax: (800) 417-2466
www.corwinpress.com

SAGE Ltd.
1 Oliver's Yard
55 City Road
London EC1Y 1SP
United Kingdom

SAGE India Pvt. Ltd.
B 1/I 1 Mohan Cooperative Industrial Area
Mathura Road, New Delhi 110 044
India

SAGE Asia-Pacific Pte. Ltd.
33 Pekin Street #02-01
Far East Square
Singapore 048763

Printed in the United States of America

Library of Congress Cataloging-in-Publication Data

Enger, Sandra K.
Assessing student understanding in science: a standards-based K–12 handbook / Sandra K. Enger and Robert E. Yager. — 2nd ed.
        p. cm.
Includes bibliographical references and index.
ISBN 978-1-4129-6992-5 (cloth)
ISBN 978-1-4129-6993-2 (pbk.)
    1. Science—Study and teaching (Elementary)—Standards—United States. 2. Science—Study and teaching (Secondary)—Standards—United States. 3. Science—Ability testing—United States—Handbooks, manuals, etc. I. Yager, Robert Eugene, 1930– II. Title.

LB1585.3.E53 2009
507.1—dc22   2009017325

This book is printed on acid-free paper.

09   10   11   12   13   10   9   8   7   6   5   4   3   2   1

Acquisitions Editor:	Cathy Hernandez
Editorial Assistant:	Sarah Bartlett
Production Editor:	Jane Haenel
Copy Editor:	Jenifer Dill
Typesetter:	C&M Digitals (P) Ltd.
Proofreader:	Cheryl Rivard
Indexer:	Maria Sosnowski
Cover and Graphic Designer:	Michael Dubowe

Contents

Preface

Since we wrote the first edition of this book, the assessment of students has continued to be a major focus of education, and assessment has become an increasingly important part of teachers' professional practice. Educators have extended responsibilities for not only setting educational goals and objectives but also for instructing and assessing in ways that help students meet these goals and objectives. State assessments bring additional pressures for students to perform well, and this also is a driver for teachers to ensure that their students are prepared. Stiggins (1994) noted that teachers make decisions about how to interact with their students at the rate of 1 out of every 2 to 3 minutes, and they base those decisions on their own assessments of student learning. Preparing students to do well in the classroom and on the measures used to make this determination are keys to making valid and reliable decisions about how to instruct and what and how to assess.

Instruction and assessment should be planned together and interconnected. Assessment practices have often focused on the use of set questions that have provided a limited number of options for student responses. While this assessment approach may be representative of some tests, teachers of science realize that students should develop deeper science understandings. Continued emphases on assessments at local, state, and national levels mean that educators must persist in changing and refining the assessments that are implemented in classrooms. Assessment can be viewed as a pathway to address the following questions:

- Should assessments tell us what students cannot do or what each student can do?
- Should assessments set targets for learning or merely sample the present curriculum?
- Should students be evaluated only on their individual work or also on their abilities to work together for the benefit of a larger group?
- How can assessments encourage and recognize inventive, imaginative responses that, although unexpected, are constructive and appropriate?
- To what extent can students evaluate data, understand concepts, demonstrate process mastery, and apply what has been learned to new situations?

- How does one assess that each student can actually do what the instruction intends for him or her to do? What evidence is used to document that a student has met the learning targets?
- What can be done to help students become successful learners?
- How can students attain the desired achievement levels?

Effective educational programs are linked to assessment schemes that help students grow, develop, and succeed, and such assessment schemes should be designed to meet the stated instructional goals and objectives or learning targets of both the teacher and the learner. *Assessing Student Understanding in Science: A Standards-Based K–12 Handbook* provides both guidelines that are based on research and examples from educators who have drawn on their own work in settings that range from kindergarten through university. *Assessing Student Understanding in Science: A Standards-Based K–12 Handbook* addresses the assessment of student performance and the establishment of criteria on which to base student progress in the six domains within science. These domains relate to concepts, processes, applications, attitude, creativity, and the nature of science.

Each domain is described, and a rationale is provided for assessing student learning in that domain. Chapter 1 provides an overview, supported by the research literature, for each of the six domains of science. In Chapter 2, assessment is set in the contexts of teaching, and in this chapter, the interconnectedness of instruction and assessment is discussed. The importance of formative assessment and feedback is emphasized in this edition, and assessment practices beyond the traditional paper-and-pencil test are included. Some examples of assessment alternatives that can be implemented include concept mapping, clinical or structured interviewing, portfolios, video recording, journaling, brainstorming, open-ended questioning, and a self-report knowledge inventory.

The context for Chapter 3 is the evaluation of teaching practice, and possible ways to examine teaching practice are identified, which include action research, video records, and journals. Instrumentation to evaluate classroom practice and surveys are components of this chapter. Samples of student and teacher forms of the science-as-inquiry surveys have been designed so that perceptions can be compared.

Rubrics and scoring guides are found in Chapter 4, which describes ideas for designing schemes to assess student work. Some examples of rubrics that have some design issues have been purposely included in this chapter to illustrate some of the difficulties in the design process. Not every rubric or scoring guide is exemplary, and rubrics should be designed to capture the most important attributes to be measured or evaluated.

Chapter 5, which is new to this edition, focuses on science notebooks and sets out some background on notebook implementation. Notebook usage has been embraced by numerous science teachers across the country, and one of the strengths of the notebook may be that it does not lend itself to a one-size-fits-all approach. Teachers and students can identify the approaches that work best for them. Whatever the implementation approach, notebooks have power in that students develop a sense of ownership through providing evidence of their personal science learning. The notebook is also a context in which teachers, students, and parents can have productive discussions about student understanding.

Chapters 6, 7, 8, and 9 set out assessment examples for multiple grade levels, but often, ideas can be modified for use at various grade levels. In these chapters, the assessment examples address the six domains of science. Chapter 6 has assessment examples that have possibilities for use across all grade levels. Chapter 7 targets Grades K through 4, and Chapter 8 has examples recommended for Grades 5 through 8. Chapter 9 presents some ideas for assessment in Grades 9 through 12.

This assessment handbook has a history to which many educators have contributed, and many of the initiatory ideas and samples were generated by Iowa teachers and students in their science classrooms. Many of these ideas and samples have undergone an evolutionary process in the hands of those who have edited these samples and ideas in attempts to clarify the thinking and communication intended by these instruments. The refinement process and the task of improving assessment practices have been guided by the assessment tenets that follow.

ASSESSMENT TENETS

- Assessment design is guided by the purpose of the assessment.
- Assessment includes multiple measures that are used to inform instruction.
- Preassessment of student understanding is vital in determining preconceptions and should be completed and documented by the teacher prior to the introduction of each new concept.
- Evidence of student learning is documented throughout the year.
- Assessment provides information about what the student can do rather than what the student cannot do.
- Assessment is viewed in terms of the growth in learning of each student.
- Students are assessed both on an individual basis and on their involvement in group work.
- Assessment tasks should be meaningful, challenging, and engaging throughout instructional activities.
- Assessment tasks are set in a real-world context and should have relevancy for the student.
- Central to any assessment scheme are process skills, concepts, attitudes, creativity, understanding the nature of science, and applications to the real world.
- Process skills and cognitive behaviors are assessed throughout learning or instructional activities rather than at the completion of a unit or chapter.
- Responsibility for assessment is shared with the students.
- Assessment is implemented throughout each lesson and includes all student activities that relate to the six domains of science.
- Outcomes of various student assessments, including student interest, direct and drive instruction within learning situations.

Acknowledgments

Corwin gratefully acknowledges the contributions of the following reviewers:

Mehmet Aydeniz, Assistant Professor of Science Education
Department of Theory and Practice in Teacher Education
The University of Tennessee
Knoxville, TN

Stephanie Blake, Chemistry and Physics Teacher
Parkview High School
Springfield, MO

Diane Callahan, Science Teacher
Fairfield Middle School
West Chester, OH

Lesa Clarkson, Assistant Professor
University of Minnesota
Minneapolis, MN

Jennifer Sue Flannagan, Director
Martinson Center for Mathematics and Science
Regent University
Virginia Beach, VA

Darleen Horton, Science Teacher
Chenoweth Elementary School
Louisville, KY

Rachel Hull, Fourth-Grade Teacher
Buffalo Elementary School
Buffalo, WV

Jane Hunn, Science Teacher
Tippecanoe Valley Middle School
Akron, IN

Michael Ice, Second-Grade Teacher
Field Elementary School
Louisville, KY

Gail Marshall, Assistant Professor of Science Education
University of West Georgia
Villa Rica, GA

Petrina McCarthy-Puhl, Forensic and Environmental Science Teacher
Robert McQueen High School
Reno, NV

Pattie Palmer, Fifth-Grade Teacher
Wynford Elementary School
McCutchenville, OH

Diana Parker, PreK Teacher
Robert Frost Elementary School
Bourbonnais, IL

Marilyn Steneken, Life Science Teacher
Sparta Middle School
Sparta, NJ

About the Authors

Sandra K. Enger is an associate professor at The University of Alabama in Huntsville (UAH) and is associate director of the Institute for Science Education. She teaches methods courses for preservice teachers in elementary science and also undergraduate and graduate assessment courses. As a professional development provider, she has worked with international teachers from Russia, the Ukraine, and multiple trans-Eurasian countries. Her consulting work includes development of assessment materials, conducting project evaluations, and participation in alignment studies. She graduated from Winona State University (MN) with a BS in science and a master's degree in biological sciences. Her PhD in science education was awarded by The University of Iowa. She has had extensive classroom experience at both the secondary and university levels.

Robert E. Yager is a professor of science education at The University of Iowa, where he also earned two graduate degrees. He has directed over 100 National Science Foundation projects and has served as chair for nearly 130 doctoral students. Yager has served as president for seven national professional organizations. He has been involved internationally with special ongoing projects in Korea, Taiwan, Japan, Turkey, and Estonia. Yager's research interests and teaching are involved with science, technology, and society (STS), especially in terms of it as an instructional reform effort and the other visions outlined in the *National Science Education Standards*. He continues to identify exemplary science programs across the world.

A Framework for Assessing Student Understanding in Science

ASSESSMENT BASED ON SIX DOMAINS OF SCIENCE

When humans use scientific knowledge and technology, global awareness becomes critical for environmental protection. As the American Association for the Advancement of Science (1990) stated in *Science for All Americans,*

> What the future holds in store for individual human beings, the nation, and the world depends largely on the wisdom with which humans use science and technology. But that, in turn, depends on the character, distribution, and effectiveness of the education that people receive. (p. vi)

Accordingly, scientific literacy has become a major goal of science education. Although there is no consensus regarding what kinds of science content are necessary for scientific literacy, a scientifically literate person is believed to be one who appreciates the strengths and limitations of science and who knows how to use scientific knowledge and scientific ways of thinking in order to live a better life and make rational social decisions.

Learning that fosters scientific literacy should promote development in the following areas:

- Students' inquiry skills and abilities
- Students' abilities to apply what is learned to new contexts
- Students' content and conceptual understanding
- Students' understanding of the nature of science

Yager and McCormack (1989) proposed that science had been viewed as a body of knowledge consisting of facts, figures, and theories, and this led to science instruction characterized by presentations of factual information. Yager and McCormack believed that science education that stressed what they called the "Knowing and Understanding" domain limited students in developing the level of scientific literacy demanded by the needs of society and the world. They first proposed that science education might be viewed in the context of five domains but later expanded these to six domains when they included the nature of science. As shown in Figure 1.1, an assessment framework for science learning and experiences to promote science literacy can be organized around six domains.

Figure 1.1   The Six Domains of Science

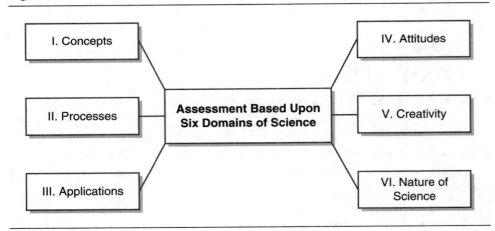

Yager (1987) noted that what was typical of science tests were questions that assessed factual, recall types of knowledge, and with grant-funded projects such as the NSF-funded Iowa Chautauqua project, he emphasized and promoted assessment in all six domains. These expectations to assess more broadly present challenges in identifying and creating assessments that measure the components in the six domains. Table 1.1 shows the foci of the six domains. If the domains are intended to support and anchor science, then instruction and experiences that provide learning opportunities in these areas are necessary for student understanding in science.

Table 1.1   What Characterizes Each of the Six Domains?

Science Domain	Domain Foci
I. Concepts (knowing and understanding)	Scientific information—facts, concepts, laws, hypotheses, and theories accepted by the scientific community
II. Processes (exploring and discovering)	Processes of science, how scientists work and think
III. Applications (using and applying)	Applications of what is learned to do science, connections to everyday life; informed decision making
IV. Attitudes (feeling and valuing)	Attitudes, sensitivity, societal issues and impacts
V. Creativity (imagining and creating)	Idea generation, designing, problem solving
VI. Nature of Science (the scientific endeavor)	History and philosophy of science; how science progresses and science knowledge and understanding develop

I. CONCEPTUAL DOMAIN

What Research Says About the Conceptual Domain

Science concepts are central to science instruction, and students' understanding of these concepts is crucial to successful teaching and learning. Millar (1989) noted that without an understanding of science concepts it would be nearly impossible for students to follow much of the public discussion of scientific results or public policy issues pertaining to science and technology. According to Thagard (1992), conceptual systems are primarily structured via either *kind* (or *is-a*) hierarchies (e.g., Tweety is a canary, which is a kind of bird, which is a kind of animal, which is a kind of organism) or *part-whole* hierarchies (e.g., a toe is part of a foot, which is part of a leg, which is part of a body). If a basic goal of science education is to help students construct an understanding of the natural world, then students' prior knowledge should be the starting point for instruction.

Assessment enters the field of view to help make determinations on where students are with respect to conceptual understanding. Students should have concrete experience with concepts before moving to abstractions, and they need opportunities to try and to do, not just to read about science. The evidence that science concepts have been learned can be seen most clearly when students can use concepts in a real-life or real-world situation (National Science Teachers Association [NSTA], 1982).

Science in the classroom has been viewed and practiced for decades as a body of knowledge or facts to be learned or absorbed by students. Classically, this occurs via the memorization of facts and concepts from a textbook. Science facts are clearly important, but to memorize facts as if their acquisition is the sole purpose of science education violates the spirit and nature of science. This issue will be addressed further in Section VI, Nature of Science Domain.

What the Concept Domain Includes

Facts, laws or principles, theories, and the internalized knowledge held by students all fall under the umbrella of the concept domain (Yager & McCormack, 1989). These are the currently accepted scientific constructs related to all of the sciences, and students may best learn these concepts through a curriculum that is conceptually sequenced for developing student understanding. Students must also experience the curriculum from conceptually sound models of assessment and instruction. Science learning should promote conceptual linkages instead of a concepts-in-isolation approach. Concept mastery is an essential aim . . . but only when a meaningful context has been established. Both *Benchmarks for Science Literacy* (American Association for the Advancement of Science [AAAS], 1993) and the *National Science Education Standards* (National Research Council [NRC], 1996) provide recommendations about content, concepts, and contexts. The American Association for the Advancement of Science (AAAS) has also published two atlases of science literacy where concept strand maps show connections across grade bands (AAAS, 2001, 2007).

II. PROCESS DOMAIN

What Research Says About the Process Domain

Science processes, often designated as *inquiry skills*, are embodied in the terms *exploring* and *investigating*. In science, the investigative processes require hands-on and minds-on activities, laboratory inquiries, and experiments that provide approaches for helping students understand scientific concepts. A study conducted by Shavelson, Baxter, and Pine (1992) found that students with experience in hands-on activities could reliably note their own progress in laboratory activities. More important, these kinds of inquiry skills are also necessary for dealing with everyday life and play a role in the development of an understanding of the natural world (Aikenhead, 1979). The contexts in which the inquiries are set are important in helping students connect the inquiry skills to their personal experiences so that students do not see the processes used in doing science as entities used *only* in science. The application of process skills in a variety of contexts also supports the development of an understanding of the nature of science. Knowledge of the process of constructing and communicating new scientific representations has the potential to yield important insights for science education (Nersessian, 1989).

What the Process Domain Includes

The process domain includes the 13 processes identified by the American Association for the Advancement of Science (1968), which provided the framework for the program in *Science: A Process Approach*. These are the generally accepted processes that scientists use as they accomplish their work, and slight variations in how these are categorized do exist. The abilities to use these process skills can be the target for instruction and assessment, but the identification of separate and distinct processes does not mean that they always occur in definable or identifiable ways. Scientists and students may use several of the science process skills in concert, and these skills may be employed during scientific investigations in ways not expected or predicted by anyone observing the investigative process. These processes and skills are embedded in knowing, doing, and thinking in science.

Process Skills Used in Science

- Observing
- Using space and time relationships
- Classifying, grouping, and organizing
- Using numbers and quantifying
- Measuring
- Communicating
- Inferring

- Predicting
- Identifying and controlling variables
- Interpreting data
- Formulating hypotheses
- Defining operationally
- Experimenting

A Brief Note on How Observation Is Theory Laden

The process of making observations may be influenced by what a person already knows about the subject or object of the observation. The prior knowledge and the conceptual framework that already exist in a person's schema influence the nature and depth of observation. This framework may or may not be accurate, but observations are made within the context of this framework. What a person can see depends on what he or she believes, and the observation is therefore theory laden (Abimbola, 1983). Personal viewpoints and creativity also play roles in any investigation, and this brings the implication that the beginning point for investigation should be based on student ideas and questions. Such ideas come from students' prior scientific knowledge, deduction, or even personal guesses and creativity. This idea that observation is theory laden may not have been overtly discussed or considered by students.

A Brief Note About Confirmatory Laboratory Work

Student laboratories and experiments can become exercises in finding the one right answer; whereas learning a protocol or procedure is often necessary in science, student experiences should move beyond this. Laboratories and investigations should provide opportunities for students to test their own ideas so that they use and develop their abilities in the process domain. Student-generated ideas can serve as the basis for the question or hypothesis that typically precedes any investigation.

A science teacher does need to play a role as the advocate of current "public concepts" (the currently accepted scientific thought) to challenge students' "private thought" (Matthews, 1994) or to persuade (Kuhn, 1962) or convince a student to appreciate the current, prevalent interpretation or explanation of natural phenomena. Group discussion for an investigation may produce the same persuading effect (Johnson & Johnson, 1983). Students who understand the role of process skills in scientific investigations may be more likely to see science as a career that has the potential to be fun and creative.

III. APPLICATION DOMAIN

What Research Says About the Application Domain

A key element in the application domain is the determination of the extent to which students can transfer and effectively use what they have learned to a new situation, especially one in their own daily lives (Grönlund, 1988). Students must demonstrate that they not only grasp the meaning of the information and processes but that they can also make applications to concrete situations that are new to them. The application domain is important because it involves students using concepts and processes not only in a familiar context but in addressing new problems. Students who can apply what they have learned to new situations provide evidence that they have an understanding of a concept.

Two major arenas where students use applications are in school and in daily life. In school, application often involves problem solving or learning new material by using knowledge and skills acquired in previous studies. In daily life, the crucial factor appears to be the ability to choose the concepts and skills pertinent and relevant for dealing with novel situations. In helping students make applications and connections among science, technology, and their personal lives, the use of current social and technological issues can assist students in seeing the need for the integration of knowledge and skills. Beginning science learning based on students' concerns in the so-called real world may be a way to diminish the learning gap between the world of school science experiences and their personal societal and technological experiences (Yager & McCormack, 1989). An issues-based approach to science learning can serve as a vehicle for engaging students in learning that is local, personal, and relevant.

Attributes of the Application Domain

- Use of critical thinking
- Use of open-ended questions
- Use of scientific processes in solving problems that occur in daily life
- Abilities to make intradisciplinary connections—integration of the sciences
- Abilities to make interdisciplinary connections—integration of science with other subjects
- Decision making related to personal health, nutrition, and lifestyle based on knowledge of scientific concepts rather than on hearsay or emotions
- Understanding and evaluation of mass media reports on scientific developments
- Application of science concepts and skills to technological problems
- Understanding of scientific and technological principles involved in common technological devices

A Brief Note About Science, Technology, and Society

Science-Technology-Society (STS) is an approach characterized by a focus on the integration of science and technology (Yager & Roy, 1993). With the STS approach, local issues provide a science context that has greater student relevancy, and the students learn concepts through activities and community action. With STS, the area in which students live can serve as the venue in which to learn science. If students live in an area with wetlands, science can be experienced in the context of the wetlands. Students might study water and habitat quality, and they could develop environmental impact studies. What are the potential benefits or detriments that exist for the community if, for example, a land developer changes the habitat to build homes needed to support economic development? In an urban setting, students might study the need to add green spaces. Students become involved in resolving local issues and proposing solutions, and this approach also pushes students to seek current information from a variety of sources and experts in various fields of study. With technology, online collaboration among students generates even greater numbers of possibilities to examine issues.

IV. ATTITUDE DOMAIN

What Research Says About the Attitude Domain

How many times have you heard people say that they were never good at science, mathematics, or some other area of study? How important is attitude anyway? Felker (1974) found that when students were induced to make positive statements about themselves, they attained more positive attitudes about themselves. Page (1958) indicated that teachers who reflected an active and personal interest in their students' progress were more likely to be successful in enhancing the personal confidence levels of students.

Attitude is very broadly used in discussing issues in science education and is often used in various contexts. Two general categories that are distinguishable are (a) attitude toward science (i.e., interest in science, attitude toward scientists, and attitude toward social responsibility in science) and (b) scientific attitude (i.e., open-mindedness, honesty, or skepticism) (Gardner, 1975). Interest in science tends to decline as students experience more science classes and progress through school. This is especially true in the middle school years when enrollment in science classes declines. Science educators need to work to retain student interest in science and need to consider changing both instruction and assessment practices to be more student-centered in order to promote ongoing interest.

The positive "I can" attitude and "I enjoy" feelings may enhance students' efforts to seek answers for their own problems and lessen their reliance on others. Students should be able to solve problems with greater independence without parent or teacher intervention. Statements such as, "Don't tell me the answer," or "I can figure it out all by myself," indicate a growing autonomy. The end result of this self-directed growth could very well be self-acceptance and responsibility for lifelong learning.

> ### Attitude Domain Attributes
>
> The attitude domain calls for experiences that support
>
> - exploration of human emotions,
> - expression of personal feelings in constructive ways,
> - decision making about personal values,
> - decision making about social and environmental issues,
> - development of more positive student attitudes toward science in general,
> - development of positive attitudes toward oneself (an "I can do it" attitude), and
> - development of sensitivity to and respect for the feelings of other people.

A Little Attitude Adjustment, According to Charles Swindoll

Although this attitude may seem beyond the scope of a science classroom, consider the words of Charles Swindoll (1994):

> The longer I live the more I realize the impact of attitude on life. Attitude to me is more important than facts. . . . I am convinced that life is 10% what happens to me and 90% how I react to it. And so it is with you. . . . We are in charge of our Attitudes.

V. CREATIVITY DOMAIN

What Research Says About the Creativity Domain

Creativity is integral to science and the scientific process and is used in generating problems and hypotheses and in developing plans of action (Hodson & Reid, 1988). Torrance (1969) defined creativity as the process of becoming sensitive to problems, deficiencies, gaps in knowledge, missing elements, and disharmonies. He also included in his description the identification of the difficulties, the search for solutions, making guesses, or formulating hypotheses about the deficiencies. The testing and retesting of these hypotheses, possibly modifying and retesting them, and finally communicating the results all relate to and rely on the creative process.

Creativity plays an integral role in the many processes of science and in doing science. Creativity is a complex construct, is difficult to assess, and rests very often in what might be called *recognizing it when you see it*. If a science educator wishes to foster a classroom that enhances students' creativity, the classroom should probably become more student-centered. Creativity is fostered and nurtured via richness in experiences. Creativity calls for openness in the classroom, acceptance of ideas, thinking outside of the box, a try-new-things approach, and a so-called go-with-the-flow approach. In fact, Csikszentmihalyi (1990, 1996) used the word *flow* as descriptive of the state in which creativity is turned on in individuals.

Studies have suggested that the work done in the laboratory rests on the ability to manipulate the objects and the instruments used. Three features of laboratory practice make the need for creative abilities paramount. First, scientists and students do not work with the natural world *as it is*; rather, they manipulate the objects of study to make them more accessible for experimentation. Second, investigators do not work with the natural world *where it is* but are instead able to bring those natural objects into an artificial or vicarious setting (e.g., the laboratory, the classroom, on a slide). Third, scientists and students do not need to study an event *only when it happens* but, rather, can cause the event to occur unnaturally when the situation demands it (Knorr-Cetina, 1981). These three characteristics of a laboratory require an imaginative, inventive mind capable of performing these investigations. These aspects of the scientific enterprise are often ignored in the traditional classroom, yet they are integral to science instruction.

Scientific experiences that can push the creative domain are likely to have some of the following attributes.

Creative Domain Attributes

The creative domain calls for experiences that promote

- visualization—production of mental images,
- generation of metaphors,
- divergent thinking,
- imagination,
- novelty—combining objects and ideas in new ways,
- open-ended questioning,
- solving problems and puzzles,
- consideration of alternative viewpoints,
- designing devices and machines,
- generation of unusual ideas,
- multiple modes of communicating results, and
- representation in various ways and modes.

VI. NATURE OF SCIENCE DOMAIN

What Research Says About the Nature of Science Domain

The endeavors undertaken by researchers in their attempts to understand the natural world can promote students' understanding of how science progresses. Science is a human endeavor that relies on reasoning, insight, energy, skill, and creativity (NRC, 1996). Honesty, values, open-mindedness, and what *Benchmarks for Science Literacy* (AAAS, 1993) denotes as *habits of mind* all play roles in scientific ways of knowing. Working with science teachers to develop their understanding of the nature of science is recommended so that they can then provide instruction that promotes their students' understanding of this construct. Preservice or inservice courses that emphasize the nature of science can result in significant gains in teacher scores on instruments designed to measure understanding of this construct (Akindehein, 1988; Barufaldi, Bethel, & Lamb, 1977). Selected journal articles, discussions, activities, curriculum projects, and media can be used to help build understanding in this domain.

In the course of human history, people have developed many interconnected and subsequently validated ideas about the physical, biological, psychological, and social worlds (AAAS, 1990). Successive generations, enabled by these ideas, have achieved more comprehensive and reliable understanding of the human species and the environment. These ideas have been developed through particular ways of observation, thought, experimentation, and validation. These ways are the bases of what is meant by *the nature of science*, and they are reflective of how science differs from other ways of knowing (AAAS, 1990).

How scientific knowledge has developed and the roles scientists have played during the process are two fundamental aspects that are important for students to know. Raising student awareness and developing an understanding of these aspects should be included in science learning. Science itself is dynamic; as witnessed by history, many ideas have come and eventually been replaced or discarded. Many science educators suggest that instruction in a science classroom should reflect the tentative nature of scientific knowledge (Lederman, 1992).

That scientific knowledge is tentative has two facets that should be expressed explicitly when working with students. First, the purpose of science is to develop a systematic knowledge in order to understand how nature behaves. Students should see science as a human endeavor in which scientific knowledge is developed by humans in an attempt to make sense of the world. Accordingly, scientific knowledge is not a truth to be discovered in the natural world but rather a man-made explanation. Second, scientific knowledge can be changed, shifted from one point of view to another, due to external social influences such as politics, economics, and culture (Kuhn, 1962). This suggests that scientific knowledge is not absolutely objective. Therefore, the understanding of the involvement of social factors in scientific development provides another focus for science education.

Science, accompanied by the power of technology, has unique characteristics that affect society. Perhaps no other human activity has ever played such a role in shaping the directions in which societies have moved. The potential to do good is very often offset by the power to cause harm, and long-term outcomes and effects are not always predictable.

An important aspect of the nature of science is related to how scientists think and work in the scientific community. Helping students to understand more of the nature of science can promote deeper understanding of what it means to *do* science. Science is often portrayed as a major intellectual pursuit of truth. Based on that expectation, many people view scientists as a group of people who are more objective and intelligent than others. Students often believe that scientists can solve problems merely based on their scientific knowledge. Science, however, is a human activity that engages real people.

In doing science, scientists often work collaboratively, and given the specializations in science and related areas, a team approach is very often needed to work on problems. Seldom do scientists work in isolation; a laboratory involves teamwork. In order to ask questions and work at finding solutions to problems, scientists must both share information and obtain information from others in their field, and most important, they must reach a consensus by virtue of discussion and persuasion and not on the basis of mere evidence. Peer review is an important component of doing science, and scientists expect to be challenged and to defend the work they have done. The work that scientists do must be replicable so that others can verify the work. Science is also a venture characterized by these competitive elements: being first to report findings, competing for research money, achieving status within the scientific community, and acquiring status for a university.

Science instruction in the classroom should attempt to portray the nature of the discipline—not simply to study the information and interpretation included in the textbook. Views currently held as so-called truths of science have changed and will continue to change throughout time. Therefore, teaching only for the retention of facts without grounding them in real-world experiences will, sooner or later, only result in the loss of these facts from memory. In an attempt to reflect the nature of science, group work, reporting findings, discussion, and reaching consensus are all parameters involved with the nature of science domain.

The Nature of Science Attributes

The nature of science domain calls for experiences that address

- the framing of questions for scientific research;
- the competitive side of scientific research;
- the methodologies used in scientific research;
- the interactions among science, technology, economy, politics, history, sociology, and philosophy;
- the ways in which teams cooperate in scientific research;
- the history of scientific ideas; and
- the ways in which science builds understanding of the natural world.

ASSESSMENT APPROACHES ALIGNED WITH THE SIX DOMAINS

Assessment approaches should include multiple measures of what students know and can do as a result of their learning experiences. The use of a multifaceted assessment approach has the potential to provide a better profile of student understanding in the six domains, and a more holistic assessment approach relates to the whole student (Raizen & Kaser, 1989). Although standardized tests may be valid in measuring knowledge of facts, they may lack validity in measuring higher-level thinking processes, investigation skills, and practical reasoning (Aikenhead, 1973; Champagne & Newell, 1992). This is not to say that standardized tests cannot and do not measure higher-level thinking processes. The deeper issue involved is one of having the assessments align with instruction and intended student outcomes.

Assessment has served as a feedback system to inform teachers about the effectiveness of classroom instruction and to inform students about how well they are learning. Information from assessments has also been used for accountability and policy decisions. The use of information from assessments influences schools, and teachers may teach to tests knowing that their teaching effectiveness may be judged by their students' scores. Thus, assessment modes can have a profound influence on teaching—often limiting classroom activities to exercises that will be on the tests (Champagne & Newell, 1992). If teachers are compelled to teach to tests, then the tests should emulate the desired student learning outcomes.

The current focus on assessments stresses the necessity to link instruction and learning with assessment. Assessment should be used as a tool for communicating expectations of the science education system to those concerned with science education (NRC, 1996). Accordingly, assessments should be embedded in the learning context to help and guide student understanding on the way toward a cumulative assessment. Having student involvement in the development of assessments and encouraging students to self-assess should empower students in taking greater responsibility for their own learning.

The shift from teacher-centered instruction to more student-centered learning may be challenging for both teachers and students. Teachers must adjust teaching practices to include a variety of instructional strategies and should consider teaching and learning environments that move students toward the goal of becoming self-directed learners. The NRC has provided some assessment examples and procedures in the *National Science Education Standards* (NRC, 1996) and in *Classroom Assessment and the National Science Education Standards* (NRC, 2001a). The standards state that the assessment process is composed of the following:

- Data use
- Data collection
- Methods to collect data
- Users of data

These four components are interrelated and are used for making decisions and taking action based on the data. The *National Science Education Standards* (NRC, 1996, 2001a) include the following assessment standards:

Standard A: Assessments must be consistent with the decisions they are intended to inform.

Standard B: Achievement and opportunity to learn science must be assessed.

Standard C: The technical quality of the data collected is well matched to the decisions and actions taken on the basis of their interpretation.

Standard D: Assessment practices must be fair.

Standard E: The inferences made from assessments about student achievement and opportunity to learn must be sound.

2

Assessment in the Contexts of Teaching

In this chapter, concurrent planning for instruction and assessment, assessment types, and methods are reviewed. Recommendations for assessments from the *National Science Education Standards* (National Research Council [NRC], 1996) are included, and ways of applying assessment standards in practice are addressed.

PLANNING INSTRUCTION AND ASSESSMENT

How do you know if students have learned what was expected? What would you accept as evidence of learning and understanding? In *My Fair Lady*, Professor Higgins exclaimed, "I think she's got it. I think she's got it," upon hearing his pupil speak properly. When do we know that our students have gotten it? Will you know it if you see or hear it? Some things are easier to recognize than others. Figure 2.1 depicts and maps some components of the interrelationships between instruction and assessment.

In today's accountability climate, educators are perhaps more cognizant of assessment practices and ways to assess student learning than ever before. Planning for instruction should also mean planning for assessment. When planning for instruction and assessment, an approach that is supported by the work of Wiggins and McTighe (2005) is one that they have articulated as "understanding by design" (UbD). A great value in this approach is that it helps clarify the kinds of understandings that we desire for students. The UbD approach can work for multiple curricular areas, and it is an approach that helps construct the bridge between the learning targets and the student outcomes.

Figure 2.1 Interrelationships Among Curriculum, Instruction, and Assessment

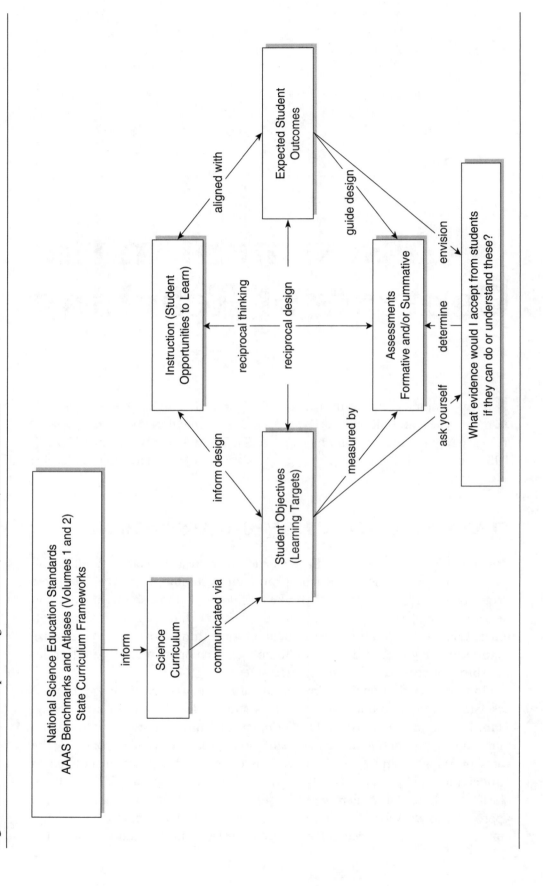

Beginning with the end in mind is an apt description of this journey. Instead of assessment being planned as an afterthought, assessment is on the planning table from the beginning.

WHAT ASSESSMENT IS

Assessment is a process of information collection, and in the context of teaching, this information is typically used to examine and describe student performance. Assessment, from the Latin *assidēre*, means *to sit beside or assist in the office of a judge*, and in current education usage, assessment also involves providing feedback to students. The ways in which assessment information is collected can range from informal to formal, and assessment information can be used in either formative or summative ways, or in some combination thereof. In the current educational climate, assessment, accountability, and making adequate yearly progress toward goals all receive scrutiny. Assessment and assessment-related issues are frequent news and media topics. These areas have been the foci of previous attention, but they have become even more politicized with the No Child Left Behind (NCLB) legislation.

Survey multiple educators about what they desire for students, and most responses will likely fall into a category that could be summarized with a statement like, "We desire that our students be the best that they can be, and we are working with them to become that." For any of us who are teachers or in the process of becoming teachers, we know from our personal experiences the complexities and nuances of classrooms. The teaching context engages you in complex multivariate environments. As teachers, we cannot control certain components in the classroom, but we can work to become the best that we can be in areas that support student learning. The educational journey on which we embark must be grounded in a quality of practice. The assessment practices used should be linked to student outcomes, and the practices should mirror the ways in which students are learning the information. With any assessment, the underlying purpose for conducting the assessment should guide the assessment design and the use of the results. Always ask yourself, "What is the purpose?"

Reminders for Planning Instruction and Assessment

The assessments should reflect the kinds of learning opportunities students have had, and assessments should also provide opportunities to move toward higher cognitive levels where students engage in assessments beyond recall and comprehension. The following strategies may be useful when planning:

- Identify big ideas and subconcepts that are relevant to the school or district framework for science. Identify possible science-related issues that could address these concepts within the community. The local relevancy that this can provide for learning science can also be used to frame assessments that involve students in using critical thinking and problem-solving skills.

- Identify the assessment strategies and approaches that can be aligned with instruction. Are there performance tasks that can be used as assessments? How can instructional tasks be modified for use as performance tasks? Sound instructional tasks also tend to be sound performance tasks.
- Decide on several major assessment tasks that students will perform during the learning process, such as projects, group discussions, demonstrations, or presentations.
- Consider instructional and assessment possibilities that provide opportunities that address attitudes, creativity, and the nature of science.
- Select assessments that align with the kinds of information needed about student performance. Multiple-choice items are efficient for checking knowledge and comprehension, but higher-order thinking skills may call for other kinds of questions.

FORMATIVE AND SUMMATIVE ASSESSMENT

Stiggins (2008) distinguished between assessment *for* and assessment *of* learning, and he noted that student involvement in the assessment process is of major importance for learning. The kinds of assessments that are integral to student learning during the instructional phases should inform both the teacher and the students about progress toward meeting instructional targets. The assessments implemented during instruction fall into the formative category and can take on a variety of formats. Assessments that are administered at the end of a unit of instruction are typically used for summative purposes, but information from summative assessments can serve a formative role when the results are used to inform and guide future instruction. Classroom assessments can be formative at some points during instruction, and the use of the same type of assessment technique implemented at a later time can serve a summative purpose. Whether the assessment is formative or summative depends on its purpose and its placement in the instructional timeline.

Formative assessment, according to Popham (2008), is a series of planned processes. While tests can be components of these processes, tests are only some of the elements in an assessment process. Popham very clearly noted that formative assessment involves a series of actions, the focus of which is to improve student learning in the classroom, and he identified formative assessment as transformative assessment. Popham further suggested that teachers should focus on learning progressions that are characterized by a sequence of skills and knowledge that are learned before a student has the capability of understanding at deeper levels. For a teacher of science, this means it is crucial to develop an understanding of the scope and sequence of science concepts; this same thinking applies to mathematics. A concept map of the differences between formative and summative assessment is set out in Figure 2.2.

Figure 2.2 Differences Between Formative and Summative Assessment

What are some differences between formative and summative assessment?

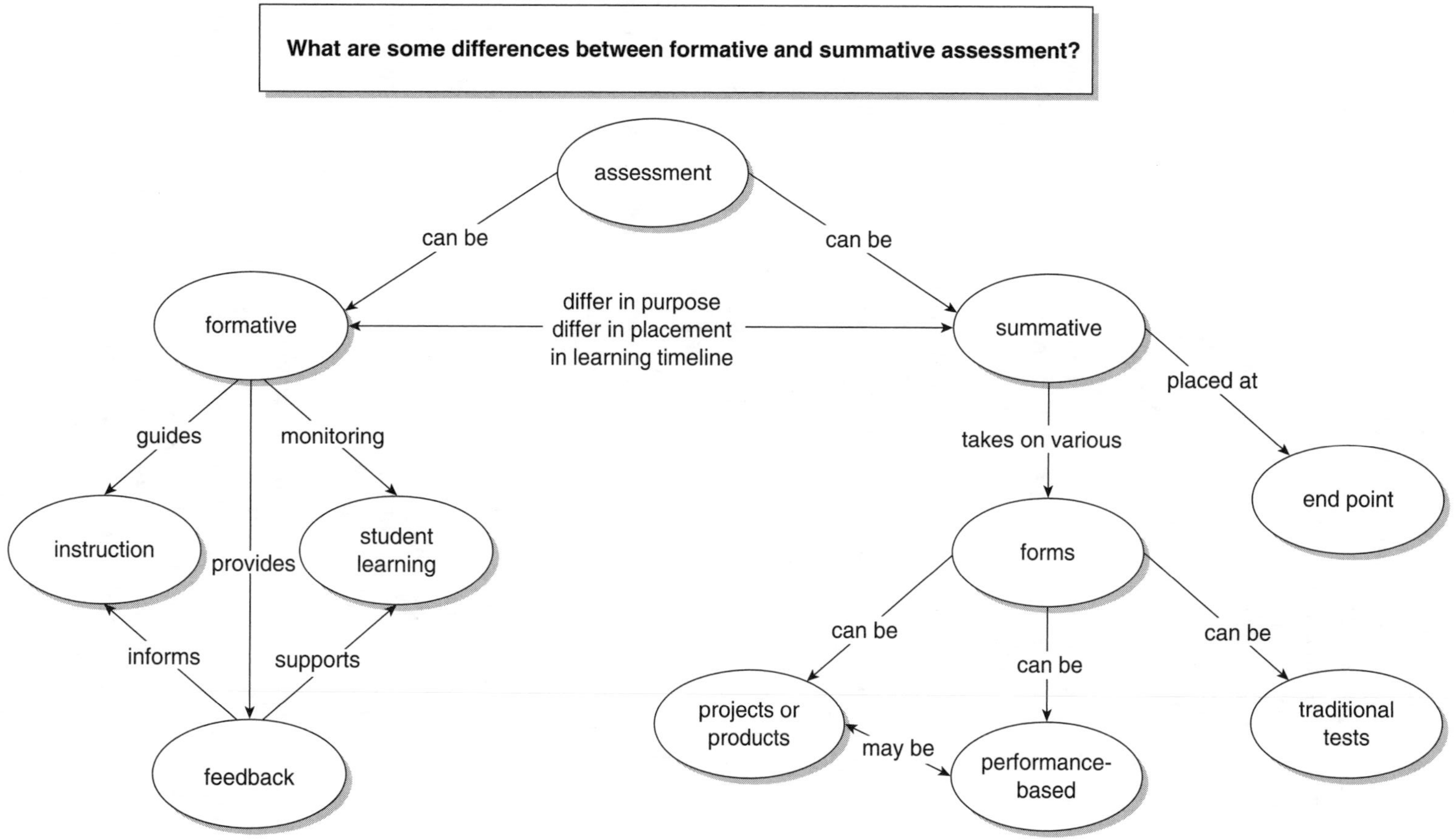

Formative Assessment and Feedback

In their study of formative assessment, Herman, Osmundson, Ayala, Schneider, and Timms (2006) noted that formative assessment is highly interactive and multidimensional. Formative assessment serves multiple functions in instruction and learning; Black and Wiliam (1998), in their meta-analytical study of assessment, found that with formative assessments in place, effect sizes ranged from 0.4 to 0.7. Also, assessments can be formative only when information from them is used to adapt and modify both teaching and learning in order to benefit students. Feedback to students plays a critical role in guiding students in making improvements. Black and Wiliam (1998) offered the following suggestions for providing feedback to students:

- Feedback to any student should be about particular qualities of his or her work, with advice on what she or he can do to improve or continue doing well. They note that comparisons with other students should be avoided.
- If formative assessment is to be productive, students should have practice with self-assessments so that they can understand the main purposes for their learning and develop an understanding of what they need to do to meet learning targets.
- Opportunities for pupils to express their understanding should be designed and included in every piece of teaching. This means that the assessment process becomes embedded and is planned along with instruction.
- Teachers should have conversations with students about their work. These conversations between students and the teacher should be thoughtful, reflective, and focused to elicit understanding. All students should participate and have opportunities to think and to express their ideas.
- Feedback on tests, seatwork, and homework should provide guidance on how each student can improve, and each student should have an opportunity to work on the suggestions for improvement.
- When feedback is provided without a grade, students tend to focus more on the feedback.

Brookhart (2008) also noted that feedback can be very powerful if it provides information that the student can use and understand. Even well-intentioned feedback can affect a student's self-efficacy, and the classroom culture must foster an atmosphere that values improvement. Opportunities for students to take action based on feedback should occur to help establish this culture, but in reality, teachers cannot *make* students focus on or learn something. When a teacher gives feedback, it is most useful to students if the feedback is given very soon after the work is completed. If feedback is provided a week or two after the work or test was completed, the feedback likely loses the impact that it might have had if given sooner. Feedback on which students can take action and improve a piece of work has more power to impact learning, but the feedback does have to be understandable and manageable for students.

Feedback can take various forms, and according to Brookhart and Nitko (2008), feedback can be

- norm-referenced when the feedback makes comparisons to other students,
- criterion-referenced when the feedback is provided according to a descriptor or criterion, or
- self-referenced when the feedback is given in relation to a student's past performance.

A general consensus on feedback is that it should be supportive in tone and should focus on the work. Work samples constructed and saved over time can provide evidence of growth. To help students recognize and realize what they are expected to understand, the processes of reviewing and critiquing anonymous work samples that illustrate a range of quality can help students internalize expectations.

Brookhart (2008) separated feedback into the components of focus, comparison, function, valence, clarity, and specificity. Is the focus on the task, the process, self-regulation, or some combination of these? Does the comparison made fall in the category of being norm-, criterion-, or self-referenced? What function does the feedback serve? Is it descriptive or evaluative? Is the feedback valence positive or negative? Generally, feedback should have a balance between positive and negative. Some teachers use a "sandwich" approach—the negative is sandwiched between two positives. Is the feedback clear and written or spoken in language that a student can understand? The level of specificity, as Brookhart noted, can range from what she called nitpicky, to just right, to too general. Sharing your thinking with students about providing feedback and modeling an action that might be taken for improvement are ways of helping students become members of the learners' club. As teachers, we can also share with students how we ourselves learn; practice does lead to success—if we help students learn what and how to practice.

Students need to be able to understand what they are supposed to be learning. Stiggins (2008) discussed writing objectives in student-friendly language to help students internalize what the objectives intend. If students were studying plant and animal cells, an objective might be to compare and contrast plant and animal cells. While this objective is already explicitly written, an even more student-friendly way to say this might be, As a student, this means that I can tell how plant and animal cells are alike and how they are different.

Seeing the Forest and the Trees

The complexities of designing and developing the instructional and assessment components for science are an opportunity for problem solving. While assessments can be designed, they cannot really be designed in isolation. When planning, design assessment opportunities that are embedded in instruction. Knowing science content, concepts, and processes—and how these are interconnected—requires a depth of understanding for the instructor. Knowing pedagogically what is developmentally appropriate for students must be considered along with what might be the most productive learning opportunities. Add to this complexity the identification of the assessment pieces and the mixture becomes even more multidimensional. Science is also challenging to understand because the intuitive explanations for events are often inaccurate, but the literature base on misconceptions can provide some insights about commonly held misperceptions in science. Being aware that misconceptions are present in science can be useful for planning instruction that is focused on accurately understanding science concepts. We have a forest, the multidimensional instruction-assessment system, in which the trees can represent things such as learning goals, learning progressions, learning opportunities, and assessments for and in support of learning.

For sources on how concepts interconnect and build, the two volumes of Project 2061's *Atlas of Science Literacy* (American Association for the Advancement of Science [AAAS], 2001, 2007) illustrate conceptual progressions in science, mathematics, and technology. Learning science is a complex activity in that learning progressions also exist in the various skills that are used in understanding and doing science. These skills may be combined and contextualized within the various areas of science. Science also requires, or even implicitly *has*, cognitive demands that would be in the highest levels of any taxonomy. Teaching and learning science is multidimensional and complex, and this presents assessment challenges. We can start with the design and use of learning progressions as guides for formative assessments. Other resources that provide insights for instruction and assessment progressions from the National Research Council (NRC) are their publications *How People Learn: Brain, Mind, Experience, and School* (NRC, 1999), *Knowing What Students Know: The Science and Design of Educational Assessment* (NRC, 2001b), and *How Students Learn: Science in the Classroom* (NRC, 2005).

Formative assessment exists for one reason: to enhance students' learning (Popham, 2008). In developing a learning progression, you can begin by designing learning tactics—the ways in which students are trying to learn something. This sounds a lot like planning lessons, but in developing a progression, assessments are planned along with the progression. As students are learning, and based upon the assessment information noted about each student, adjustments are made to the classroom instruction. This moves on to what Popham called *Level 2 formative assessment*, in which not only teacher-dictated adjustments are made in learning tactics but student-determined adjustments are also included. Will all students take ownership and responsibility for how and what is learned? A good question! A way to determine this is to actually try this out in a classroom. A golden opportunity for action research is there for the teacher-researcher. Change tends to be a slow process, and this may be a change in classroom practice for the teacher and also for the students. If we want students to take personal responsibility and ownership of learning, this could be a way to initiate these actions.

ASSESSMENT TYPES AND APPROACHES

Assessment options are numerous, and those looking to expand classroom assessment practices will find many resources available in print and online. The assessment of student performance is likely to include traditional classroom tests and standardized tests available from test development companies.

Current emphases are focused on formative assessments and also on performance assessments, and multiple measures that are intended to capture and provide a better picture of what students know and can do seem very much the trend for classroom assessments.

As instructors, our content selections for science gain input from multiple sources, including national and state standards, district or local curricular documents, and state and national assessments. Assessments influence the science curricula, and science practice materials and released items are available to inform districts, teachers, and students about targeted content and item formats. This information can be useful for preparing students with practice in test-taking skills. Also, familiarity and practice with skills like reading tabled information and graphs can support students in content areas beyond science. Assessment blueprints, tables of specification, or eligible content can be accessed via the state's Web site. Sample items and released items or tests are available, and some states also post lessons and activities that can be used in various ways to support instruction and assessment.

Most, if not all, states administer standardized tests such as Terra Nova, Iowa Tests of Basic Skills, Iowa Tests of Educational Development, Metropolitan Achievement Tests, or the Stanford Achievement Test, and the results from these tests can be used to identify areas of strengths and areas in need of improvement. Data from these assessments can also be used in formative ways to guide instruction with future students. The test results can also provide information for comparative uses across classrooms, schools, districts, and states. States usually have state assessments that are aligned with their curricular frameworks, and these state assessments are administered at selected grade levels. Test results can inform science programs, and the results can also identify target areas for professional development. The assessment system information can be useful in identifying strengths and areas in need of improvement.

For tests like Terra Nova, Iowa Tests of Basic Skills, Iowa Tests of Educational Development, the Stanford Achievement Test, or the Metropolitan Achievement Tests, you also can find the assessment blueprints or tables of specifications. This information is likely available from your district's data collection office, and practice tests and preparation materials are typically available from the test publisher. For these tests, you may find no or few released test items, and a reason for this is the time and expense necessary for test development. Also, these tests are used for comparative purposes, where student test scores are compared with the group used to norm the test. If tests are released or security is breached, valid inferences cannot be made. Usually, states will release some practice items from their state tests; when tests are developed, the number of items piloted is greater than the number of items needed for state assessment and a set of items is released for both practice and other reasons. When states make the decision to release their state-specific tests, this represents a major financial decision.

Although traditional assessments do provide information, a concern exists that the more traditional standardized tests cannot represent a complete picture of student performance in science (Champagne & Newell, 1992; Pierce & O'Malley, 1992). Few would deny that standardized tests are indicators of some facets of student achievement, especially in large-scale testing, and that they should not necessarily be discarded from assessment options. Information from large studies, such as the Third International Mathematics and Science Study (TIMSS), now the Trends in International Science and Mathematics Study, and the National Assessment of Educational Progress (NAEP), the Nation's Report Card, do provide longitudinal assessment information for comparative purposes. Results from these assessments provide contexts for discussions on the nature of programs and what students are expected to know and be able to do.

Traditional assessments typically fall in the category of paper-and-pencil types, which are the true-false, fill-in-the-blank, multiple-choice, and short-answer-type tests. When science instruction becomes more student-centered and more constructivist in orientation, a wider range of assessments may be needed to capture student performances. If learning is viewed as an active process (Ausubel, 1968) and if scientific knowledge becomes meaningful to students when they have opportunities to learn science through inquiry (NRC, 1996), then the assessments used should reflect these experiences. With an interactive picture of student learning in mind, current science educators argue that student performance in science should be assessed from various facets of learning in the different contexts of teaching. Performance assessments should be considered for use in an effort to provide multiple measures of student learning outcomes because performance assessments can more closely parallel the learning experiences. The *National Science Education Standards* (NRC, 1996) have set out assessment standards that can be used to guide assessment practices.

Initiated in 2003, the Program for International Student Assessment (PISA) is a system of international assessments that measures the capabilities of 15-year-old students in reading literacy, mathematics literacy, and science literacy every three years. The U.S. Department of Education's National Center for Education Statistics (NCES) and the Institute for Education Sciences (IES) have score reports and released items that can be accessed via their Web site (http://www.nces.ed.gov/). From this site, you can access NAEP, PISA, and TIMMS information. NAEP, for example, has a question tool where you can select items and compile an assessment that could be used either for assessment or as practice for your students. More information about using NAEP items is provided in later chapters.

Standardized Assessments

What Standardized *Means in Assessment*

An assessment is standardized in the sense that the administration, scoring, and apparatus are fixed by the assessment developers so that the test may be administered and scored by different examiners in different settings and achieve comparable results across all examinees. A fixed set of test items designed to measure a clearly defined domain and norms characterize standardized tests (Grönlund & Linn, 1990). A standardized test is often administered in a multiple-choice format, and as in the selection of any assessment, various factors play roles in the selection process. For large-scale assessments, a multiple-choice format is often used, and inherent in this format are certain advantages and disadvantages (Table 2.1):

Table 2.1 Advantages and Disadvantages of Multiple-Choice Tests

Advantages	Disadvantages
• Less expensive than some other forms of testing for large-scale testing • Easier to administer to large numbers of students • Makes ranking individuals easier • Can be used to make comparisons across locations, cities, states, and nations • Can sample a wide variety of learning targets • Fast tools to test student knowledge of specific content • Can assess higher-order skills • Can be used diagnostically to improve instruction	• Expensive to develop and norm • Used to rank schools based on test information • May be used to judge schools on student performance on one test • May be used to judge teachers on student performance on one test • Can fail to assess higher-order skills • Provides just one measure of student learning • Constructed upon the assumption that knowledge can be represented by an accumulation of bits of information and that there is one right answer • Encourages teaching to the test, which narrows the curriculum

Classroom teachers do use standard procedures in assessing their students in that attention is given to assessment design, administration, and scoring procedures. The process of standardization employed by various test development companies or businesses is extensive, very in-depth, highly structured, and based on psychometric models.

Alternatives in Assessment

What Alternative Means in Assessment

Alternative assessments are usually those assessments that are alternatives to the traditional paper-and-pencil test and are not the commercial standardized multiple-choice test. Alternative assessments tend to be criterion-referenced instead of being norm-referenced. Alternative assessments can include teacher observations of student performance, student interviews, student self-assessments, presentations, projects, concept maps, portfolios, and any of a variety of formats used to provide evidence that learning is occurring. Alternatives in assessment options provide a wider range of opportunities for students to represent what they are able to do to meet learning targets. A downside of this range for opportunities to generate responses is that subjectivity in evaluation of responses can lead to greater score variability.

Alternative assessment allows for the evaluation of a wider variety of learning targets, and instructional time can be focused on developing a wider range of student skills and abilities. Furthermore, it is possible to begin to profile a student's intellectual development over time. A major weakness of alternative assessment, at least for large-scale testing, is the difficulty in achieving consistency in interpretation of assessment results (Champagne & Newell,

1992). Alternative assessments are also typically more time-consuming to administer. In addition, scoring alternative assessments usually requires a team of raters, which adds to the expense of conducting an alternative assessment. Alternative assessments are also subject to questions of reliability and validity, and even with trained raters, subjective judgments are made. Nitko and Brookhart (2007) noted that reliability coefficients for standardized multiple-choice achievement are typically in the 0.85 to 0.95 range, open-ended paper-and-pencil tests are typically in the 0.65 to 0.80 range, and the reliability of portfolio scores typically falls within the 0.40 to 0.60 range.

Even if assessments vary in reliability, a recommendation for classroom teachers would still be to use multiple measures to assess the cognitive development and academic progress of students. Integration of assessment with instruction, assessment of learning processes and higher-order thinking skills, and a collaborative approach to assessment enable teachers and students to interact in the teaching-learning process. No one measure is likely to capture total student performance. But how we teach science and how we assess student learning in science should map onto each other. Assessment practice should mirror the learning opportunities that students have had in the science classroom.

What Authentic *Assessment Means*

Authentic assessment, as viewed by Pierce and O'Malley (1992), is authentic because it is based on activities that represent actual progress toward instructional goals and because it reflects tasks typical of classrooms and real-life settings. Authentic assessment has also been defined as *how well the assessment represents a real-world task* (Wiggins, 1989a, 1989b, 1993). The most authentic assessment of whether a person can drive a car is to actually have the person drive that car. Wiggins (1998) noted that in order to be authentic, tasks, problems, or projects should be realistic, require judgment and innovation, have students actually complete the doing, be aligned with the process in the real world, possess complexity, and provide opportunities for performance refinement. To reiterate, authentic assessments should emphasize application, focus on direct assessment, use realistic problems, and encourage open-ended thinking (Baron, 1991; Horvath, 1991; Jones, 1994, as cited in Nitko & Brookhart, 2007).

What Embedded *Assessment Means*

Embedded assessment is assessment that is interwoven within the context of learning. Assessment is embedded in instruction and in the classroom opportunities to learn, and with this in mind, assessment and instruction are planned in concert. Assessments are planned so that as students are learning, both the teacher and students know that they are "getting it." Assessments are planned with anticipated student outcomes in mind, and a teacher should always try to envision what kinds of evidence indicate that a student really understands. Opportunities to assess are built right into instruction.

Sloane, Wilson, and Samson (1996) supported the model of envisioning assessment as a system, and within this system, certain components must function together for the system to operate. An assessment system involves the design and development of assessments, the implementation of the assessments, and the products produced—along with how information from these products is used to inform various system components or stakeholders. An assessment system, however, cannot have productive outcomes if it does not have interconnectivity with instruction.

In projects led by Gallagher (1999), embedded assessment was reported to have the following impacts in classrooms:

- Improved quality interactions between teachers and students
- Improved classroom environment, including diminished behavior problems
- Increased student engagement in learning
- More positive interactions among students
- Increased understanding of science and science achievement
- More positive attitudes toward science and science classes

Performance Assessment

What Performance *Means in Assessment*

Performance assessment is a process in which work assignments or tasks are used to obtain information about how well a student has learned (Nitko, 1996; Nitko & Brookhart, 2007). Performance assessments may be alternative assessments, and the actual type and format for the performance may vary. Often, some type of task is involved, and short-term and long-term projects, whether individual or group, could be considered performance assessments. A wide range of assessments could be subsumed in performance assessment. Performance assessment typically involves setting criteria for which a student demonstrates specific skills and competencies, and a rubric or scoring guide is designed to communicate varying levels of performance. Levels of student performance on the assessment are evaluated on the criteria developed and communicated in the rubric (Table 2.2).

Table 2.2   Advantages and Disadvantages of Performance Assessment

Advantages	Disadvantages
• Provides opportunities for applications of inquiry skills • Provides opportunities to practice with open-ended questions • Promotes opportunities for critical thinking • Provides more direct evidence of what students can do versus indirect evidence from traditional tests • Promotes opportunities for creativity	• Takes time for students to complete • Poses tasks that are challenging for some students to complete • Calls for subjectivity in assessing performance • Allows variation in performances and is open-ended in nature • Assesses performance that is task specific • Can involve time and group management issues • Requires time to design, implement, and evaluate student performance • Poses challenges in the design of quality tasks and rubrics

Portfolios in Assessment

What Portfolios Can Mean

In the context of assessment, a portfolio is described as a limited collection of student work that is used either to present best work(s) or to demonstrate a student's educational growth over time (Nitko, 1996; Nitko & Brookhart, 2007). This collection is driven by the purpose for which the portfolio collection was assembled. Portfolios may be used to showcase best work, to provide documentation of long-term projects, and in instances where a purposeful collection provides support and enhancement that cannot be presented in other ways. Pierce and O'Malley (1992) noted that portfolio assessment provided an approach for combining the information from both alternative and standardized assessments and that opportunities for student reflection and self-monitoring were supported by a portfolio approach. As Nitko (1996) so aptly noted, a portfolio is not a scrapbook, nor is it a "dumping ground" for all work. With current technology, e-portfolios are realities and portfolio management is supported by the abilities to edit and revise portfolio content. The portability of a portfolio is greatly enhanced with technology, and a wider audience can view and assess the portfolio. Perhaps this might be viewed as negative by some, but an e-portfolio is more easily updated and revised to showcase a student's work (Table 2.3). Managing an e-portfolio moves the portfolio into being a more dynamic and purposeful collection of work.

Table 2.3 Advantages and Disadvantages of Portfolio Assessment

Advantages	Disadvantages
• Provides student practice in decision making in content selection • Supports student practice in organization • Supports student practice in communication of work • Develops student self-assessment strategies • Provides more opportunities for students to address personal learning styles • Provides student opportunities for creative expression • Promotes student reflection on personal improvement • Promotes student responsibility for maintenance of work history • Promotes communication among student, teacher, and parent or guardian • E-portfolios support technology skill development	• Requires storage of numerous student portfolios (for paper format) • Requires decisions about the kinds of works to include • Requires decisions on how to assess portfolio contents • Requires time to monitor and assess student portfolios • Requires updating portfolios— what to keep and what to weed out (e-portfolios may ameliorate this at some level) • Requires moving portfolios from one grade level to the next; student ownership of the portfolio and responsibility for organization should be encouraged (e-portfolios may assist here with portability and management)

Portfolio Content Possibilities

The content that could be included in a portfolio is extensive (see Table 2.4), but decisions about what to include should always be guided by purpose: how

Table 2.4   Portfolio Content Possibilities

Projects, Notes, Class Work	Various Writing Samples	Journal-Type Entries	Evidence About Self	Media Usage
• Individual and team projects • Class notes, research notes, outlines • Assignments • Assessments • Graphic organizers • Best works	• Self-evaluations • Position or issue papers • Works in progress: first to last drafts • Lab reports • Creative works—poetry, essays, songs, and stories • Critiques and reviews • Letters	• Lab logs • Multiple formats • Free-write, reflective, prompt-specific comments • Reading response, science notebooks • Wishes, ideas, thoughts, reactions • "What I understand" • "Questions I have"	• Evidence of effort • Teacher and parent notes • Newspaper or media reports • Notes on work from peers and others • Creative endeavors • Community involvement, service work • Areas of greatest improvement	• Audiotapes • Digital media or video recordings • Multimedia work • Podcasts • Web pages • Photos • Artwork models • Computer programs • Best works

the contents will provide support for and evidence of student learning. The portfolio can become a comprehensive form of assessment and documentation of student growth. Student ownership of the portfolio and responsibility for organization should be fostered and encouraged.

Alternative Assessment Implementation Strategies

In this section, some suggestions are offered for strategies available to help implement alternative assessments. These include observations, clinical interviewing, concept mapping, journaling, brainstorming, open-ended questioning, and self-knowledge reporting.

Making and Using Observations

Teachers typically observe students in their classes; these observations are one way to assess student abilities in science. Teachers may not consider the act of observing as a form of assessment, but instructional adjustments are frequently made based upon our observations. Because observations are often filed in a mental notebook rather than being written down, observations serve as initiators of formative assessments. If documentation is desired, anecdotal notes can be made either on paper or with the use of available technologies. Checklists can be developed to keep records of performance of skills or other kinds of learning where it is relevant to know if someone can or cannot do something. When assessment focuses on the totality of student performance in the classroom, a productive way to evaluate student work in the laboratory or in the classroom is for the teacher to keep observational records. Direct observation and recording of those observations will give a better account of what students are capable of doing as they work in typical situations. Observation checklists that streamline some of the paperwork can be developed. A generic checklist can be developed with some spaces left for specifics.

Observations can also inform the teacher about which students know how to use the library or media center, take notes, locate resources, or use the computer, other electronic devices, and the Internet. Collaborative and cooperative work and affective components are also important for student success. If assessment is concerned with evaluating the whole student, observation can play a key role in that evaluation; observations can provide teachers with continued guidance in helping students attain maximal success in the classroom. Observational assessments can provide a comprehensive picture of student participation and involvement in the classroom. Share the kinds of possibilities noted in Table 2.5 with students so they realize that their demeanor is important in the classroom and in the workplace and community. Some examples of student behaviors that could be included as observational evidence are listed, but the generation of possibilities would also be a productive activity for students.

Table 2.5   Possibilities for Observational Notes

Examples of Student Behaviors	Evidence Observed	Date
• Has a positive attitude • Is prepared for class • Is receptive to ideas • Is receptive to feedback • Makes responsible use of time • Meets deadlines • Shows curiosity • Has an investigative spirit • Is an independent learner • Asks quality questions • Contributes suggestions • Is a productive participant • Shares group responsibility • Contributes to group work • Respects others • Respects class and lab facilities	*(Notes: The observer could set this up to make notes here. A set of descriptors could be developed. This could also be edited to be used as a student self-evaluation piece.)*	

Time Concerns

In life, there is never enough time to do all of the things that a person wishes to do. Many of the recommendations for changing assessment practice do require time, and if time could be created, the answer for managing assessment might be easier to address. Ask yourself this question: Do the assessment practices in my classroom give the best evidence of what the students know and are able to do? If the answer is yes, you won't be likely to change what is being done. If the answer is no, then perhaps some of the assessment alternatives should be considered and implemented. If your approach to teaching is one of always having room for improvement, you will likely try new methods and ways to accomplish things, including assessment practices.

Web Tools

Having technology at our fingertips has transformed how we communicate and collaborate. One of the first things we likely do if we have a question is to search the Internet to find answers. Our current students are, for the most part, technologically savvy. Richardson (2009) noted that the technology "toolbox" includes technologies that publish, manage information, and support content sharing. Blogs are everywhere, and teachers and students are blogging in classrooms around the world. Blogs can support conversations among students and between students and the teacher. The security of students is a priority, and being apprised of school policies should be a first step. Again, purpose would drive the kinds of assessment information that could be collected. The possibilities appear myriad. Podcasts, wikis, and Real Simple Syndication (RSS) are but a few approaches to incorporating technology for instruction, learning, and assessment in the classroom. Some software, like Audacity, is currently available for free download, and other types of programs are also available at no cost. Our students live in a world where they are immersed in the use of technology, and these modes can be infused in classroom instruction and assessment practices.

Concept Mapping

A concept map is a two-dimensional, graphic representation of an understanding of some domain (Novak, 1998; Novak & Gowin, 1984). Concepts related to the domain of interest are circled and linked to other concepts to show connections. Linking words indicative of the relationship are added to the lines, and a hierarchy is typically present. If the linking words are not included, the basis for linkage must be inferred; with no linking words, just a web is created. Concept mapping can be used in combination with open-ended questions to provide evidence about what students know and understand. Concept maps provide a graphic depiction of student understanding, and concept mapping is a form of graphic organization. We are not born knowing how to concept map, so before we expect students to used this mode of representation, a few practice sessions can be devoted to instruction in mapping. Figure 2.3 shows the process of concept mapping and also provides a few ideas for getting started with the process.

A recommendation for learning how to concept map is to begin the process with a topic with which students have some background knowledge. Students usually know something about pets, and in the map that follows, several questions were used to focus and initiate the mapping process. Students can work in groups to develop a map; this is a productive way to initiate mapping. Figure 2.4 shows a map that was completed using Inspiration® software, and use of software does aid in map portability. Hand-constructed maps can also be photographed for portability, or maps could be scanned if issues of size are considered in map production or reproduction.

In a classroom doing group map construction, a recommendation is to use chart paper and markers. The key words to be used in mapping can be set out by the teacher or generated by students. When beginning, keep the number of words manageable; while it depends on the age of students and the complexity of the content, 10 to 12 words as initiators usually work. Students can add extra words to show what they know, and of course, linking words are added by students. The 10 to 12 words that are to be included on every map can be written on sticky notes

Figure 2.3 A Graphic Representation of Concept Mapping

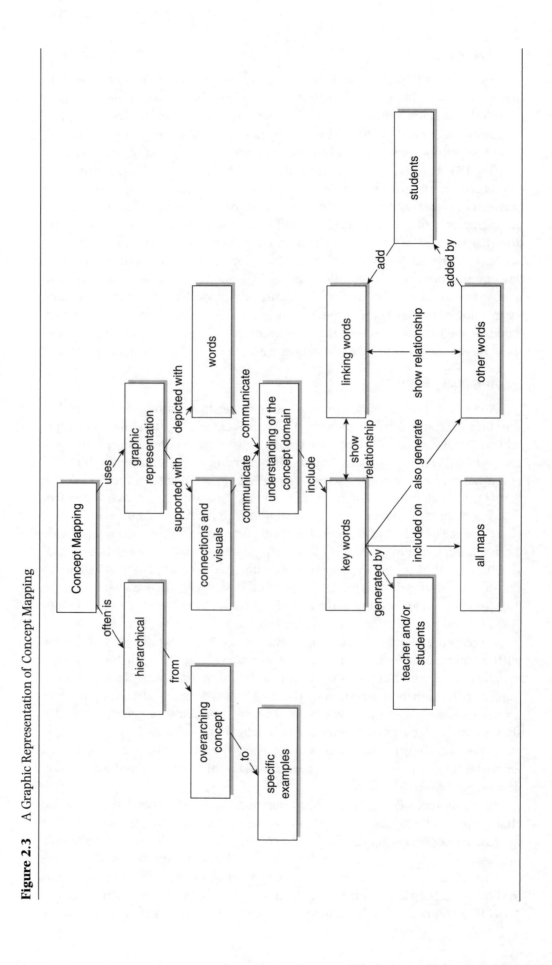

Figure 2.4 Introduce Concept Mapping With Topics Known to Students

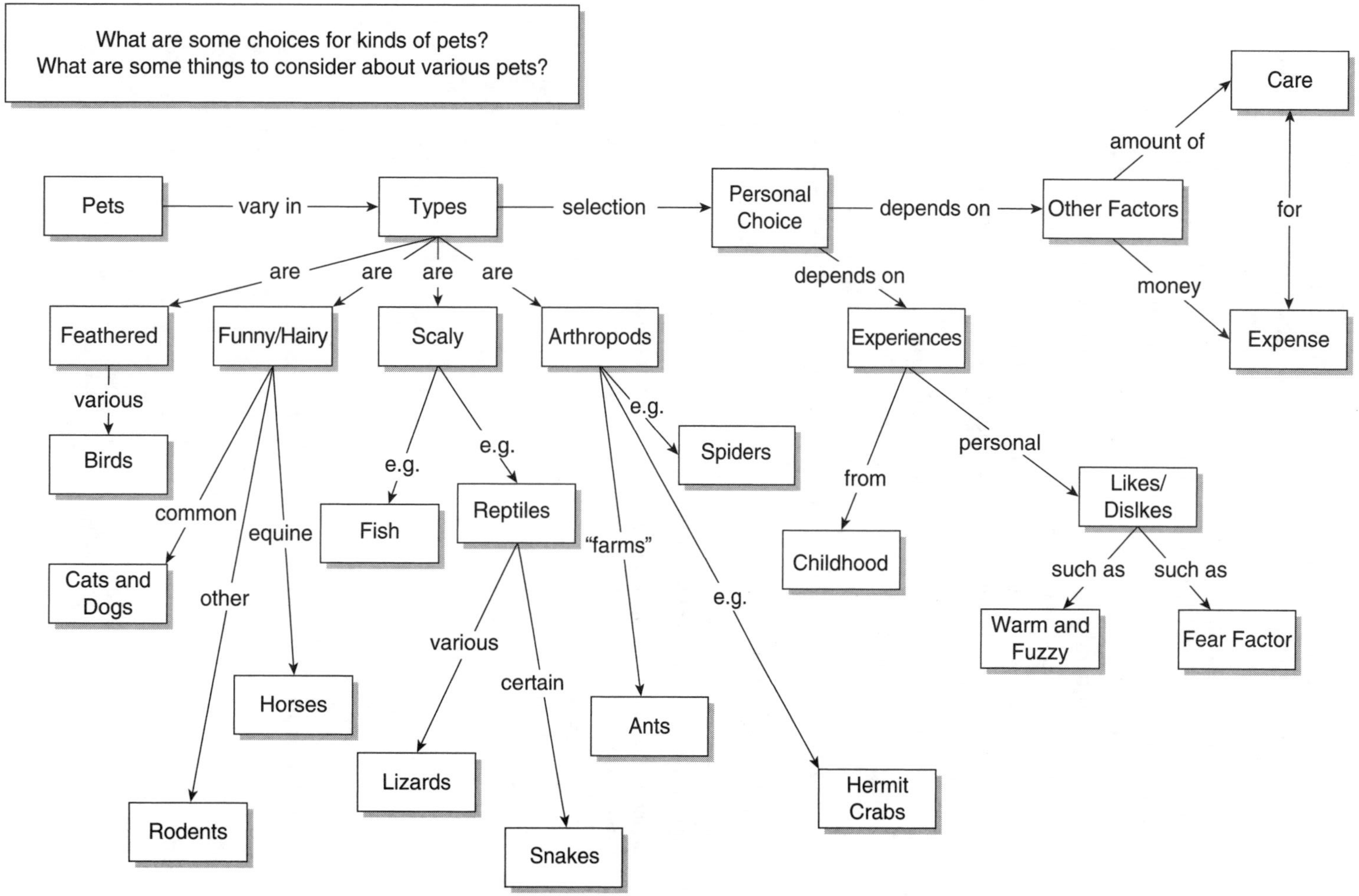

What are some choices for kinds of pets?
What are some things to consider about various pets?

or slips of paper so that the words can be positioned and repositioned before final organizational decisions are made. When students are working in small groups, they discuss where, how, and why the words should be placed and organized on their map. Students learn from one another and provide reasons for how their map represents knowledge and understanding "under construction."

As students are working on maps, the instructor can circulate among groups and ask students to talk about the organization and connections on their maps. Asking students to explain and clarify their thinking in this way is a very nonthreatening type of formative assessment. Maps can be used as both preassessments and postassessments, and they can be updated throughout instruction. The use of concept maps in these ways also represents an example of either embedded assessment or ongoing assessment.

Having student groups post their maps around the room for viewing conveys an important message: Seldom are the maps superimposable. Knowledge construction and representation are idiosyncratic to the groups, and going further, so it is for each student.

Student groups can complete a gallery walk and use sticky notes to write and post feedback on the maps—"sticky note blogging."

Concept maps could also be initiated from text, such as in the water cycle example shown in Figure 2.5. The map is generated from the sentences.

Figure 2.5   A Concept Map Initiated From Concept Statements

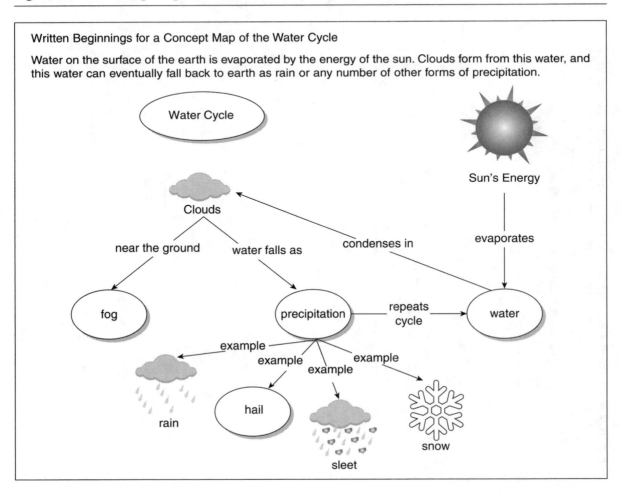

Written Beginnings for a Concept Map of the Water Cycle

Water on the surface of the earth is evaporated by the energy of the sun. Clouds form from this water, and this water can eventually fall back to earth as rain or any number of other forms of precipitation.

Alternatively, students could create concept maps first and then write paragraphs to support their maps. Combining the graphic and written works provides practice in using various modalities.

The Florida Institute for Human and Machine Cognition (IHMC) has extensive information and research background on concept mapping. The IHMC concept mapping tools, Cmap, can be accessed online (http://cmap.ihmc.us/conceptmap.html). Various other software packages for producing maps and graphic representations are also available.

Structured or Clinical Interviewing

A structured or clinical interview follows a protocol for which interview questions have been developed, and these same questions are asked of all persons interviewed. Jean Piaget developed the clinical interview, and it has been adopted by researchers concerned with the prior knowledge, alternative frameworks, or misconceptions held by students and can be used with students after concept mapping (Novak & Gowin, 1984). The interview allows the teacher to orally interact with the student and to probe for further clarification of the student's concept map. This can facilitate assessing what the student may be unable to communicate via the written words. Focus groups could also be used instead of individual interviews. Again, purpose should drive the choice.

If teachers follow some simple guidelines, the interview can be a powerful tool in establishing students' prior conceptual frameworks and evaluating student learning at any point in instruction. Remember that if the interview is to be audio or video recorded, you may need permission to do so; school districts typically have policies delineated. The time required to conduct interviews may limit how often this technique is used. However, the ideas underlying the interview are valid for incorporating into questioning strategies for a class.

Here are eight ideas for designing and conducting the structured or clinical interview:

1. Plan the interview. Where possible, use concrete objects, situations, or pictures to provide a context for the interview.

2. Design questions to find out what the student knows about the issue or the concepts involved. Avoid yes-or-no questions. Do not ask questions designed to aid student understanding of the issue. Let the student talk without fear of right or wrong. Have the student describe, predict, or explain. Word questions in student-friendly language. Establish rapport.

3. Sequence the questions from the easier to the more difficult ones. It is important to establish student confidence so the student does not become nervous and uncomfortable during the interview. The real goal is to find out what students are thinking. Again, rapport is important.

4. Students should be given adequate time to answer the questions. If the student fails to respond to the initial question or says he or she does not understand the question, rephrase the question so the student has an opportunity to respond.

5. Whether it is the first time or the 10th time for the teacher or the student, it is important to convey a relaxed presence. Interviewers should project the image that they are human and that they do not know all the answers. Rapport, rapport, rapport!

6. In any sequential interview, it is important to refer to prior interviews or relevant intervening instruction, and the student's ideas should be used. This gives the student a frame of reference from which to respond, makes the conversation roll, and gives evidence of teacher acceptance.

7. The language of the student should be used when rephrasing questions or probing further. To insist on the "right" word or pronunciation can be confusing and inhibit fuller expression of concepts and propositions. Also, students will sometimes use the wrong label (word) for the right concepts. (For example, students will often say that the earth is shaped like a circle rather than a sphere.) When this occurs, an explanation can be asked for and the correct concept label supplied by the student.

8. Last, interviews should end on a positive note. Cooperation, manners, and answers to questions by the student can be noted; any questions the student might have can be answered. In any case, the interview should end with feelings that will make future interviews an experience to be welcomed and anticipated (Novak & Gowin, 1984).

Video Recording

Video recording can provide an opportunity for self-assessment of teaching practices or assessment of student performance. Audio recording can also be used with students to help them view their own abilities and knowledge in any particular area. Simply setting up the camera and letting it run during class can provide valuable information for instructional improvement. The recordings can be evaluated with an assessment instrument at a later time, but communication about intent is crucial. If the recordings are to be used for research purposes, then parental or guardian consent is required and policies must be followed. With current technology, recording, uploading, and editing functions are quite user-friendly.

Journaling

A journal is a diary-like series of writings, or drawings, or both. The student should have a separate book or folder for the journal. Entries can be responses to an instructor's questions or statements, feelings about an activity, or can answer the question, "What did I learn today?" The focus on writing is a strength of journal implementation. Students gain practice in their writing skills while communicating their ideas on paper. Journals can take on any variety of formats driven by the intended use of the journal, and creativity can be facilitated with variety in format. Here again, incorporation of technology can move this to a digital format.

The list of positive aspects of student journals is lengthy. For example, connection of knowledge between subject areas can occur when journals are used. Teachers are better able to respond to individual questions as the journals are read at various times, and a teacher-student dialogue can be established. Long-term improvements within the course and teaching methodology can result in response to insights gained from students.

Brainstorming

Brainstorming can be used as a preassessment strategy to determine what students know about a question or issue they will be investigating. Students could first be asked individually to write down all they know about the topic. After this preliminary session, a class or group brainstorming session could occur. The teacher could, at a later point, use the papers and notes from the initial session to document and assess growth in that area. When brainstorming serves as an introductory instructional activity, questions are formulated to discover what students know in general that can serve to provide an assessment of what they do not know in the topic area. Brainstorming can also serve as the prelude to concept mapping.

To be certain that every student has a preinstruction assessment, ask each of the students in the class to take a moment to record his or her thoughts, either in a science notebook or on a separate sheet of paper, before the entire-class conversation begins. The exercise of writing before students discuss can also lead to a more engaged class discussion since the students will have focused on the topic and considered what they want to say.

At the end of each lesson and at the end of a unit of study, teachers can ask each student to write down what he or she learned. Students can also be asked to write down the questions they still have. If the information collected from the different periods of brainstorming is carefully documented and analyzed, it is likely that teachers and students will note growth in understanding from the learning experiences. These suggestions represent ways in which assessment can become embedded.

Open-Ended Questioning

Open-ended questions invite multiple solutions and multiple ways to arrive at solutions. Open-ended problems assist students in the development of problem-solving skills and serve to promote creative and divergent thinking. These questions move students away from trying to guess what the teacher wants and away from the prevalent notion that only one right answer exists for all problems. An open-ended assessment approach becomes less of a guessing game and more of a preparation for the real world where multiple solutions are not only possible but often necessary.

Develop open-ended questions by using the following strategies:

- Identify the big ideas(s) and subconcepts to be assessed.
- Determine the cognitive skills to be assessed.
- Design a sample prompt (scenario, story line, etc.) to engage students in thinking and elaboration.
- Write your own sample answer to address this prompt.
- Pilot the open-ended scenarios with students.

- Compare the student responses with your expected outcomes. Rewrite and revise based on student responses.
- Draft a scoring rubric for student responses. Introduce students to rubrics and let them assist with the design and scoring process. When possible, share the rubric with students prior to the assessment so students have a context in which to frame a response. As a pool of open-ended questions is developed, save and share exemplary responses with students so students become aware of expectations. Although this means continued assessment development, the assessment itself can become an instructional tool. Any of these suggestions could also be developed for an online format.

Self-Report Knowledge Inventory

For a preassessment of student knowledge in an area to be studied, students can be asked to rate their knowledge of concepts and skills on a five-point scale. Students select a descriptor of their self-perceived level of understanding in the area of instruction. The instructor sets out some concepts, related ideas, or relevant vocabulary on which instruction will be centered. This may also help to activate prior knowledge. A lesson on ecosystems might focus on trophic levels, and the words set out in the following frame might be used.

Sample Self-Report Knowledge Inventory:
What I Think I Know About Ecosystems

Use the following statements to indicate what you know about the concepts or words related to ecosystems.

1. I have never heard of this.

2. I have heard of this but do not understand what this is.

3. I think I somewhat understand what this is.

4. I know and understand what this is.

5. I know and understand this well enough to explain this to another student.

_____ trophic levels	_____ food chain
_____ producer	_____ food web
_____ consumer	_____ food pyramid
_____ producer	_____ biomass
_____ decomposer	_____ energy flow

Teachers can use student responses to direct or guide instruction so students can reach Level 4 or 5. This could be given as a preassessment to activate prior knowledge or as a background knowledge probe (Angelo & Cross, 1993). This kind of assessment could also be included as a kind of journal entry where students could note their changes in understanding as instruction proceeds. The ability of students to self-evaluate is an important goal of education. The time needed to design, administer, and assess with this strategy is minimal, and the results have been found to be valid and reliable (Tamir & Amir, 1981).

NATIONAL SCIENCE EDUCATION ASSESSMENT STANDARDS

The assessment standards provide criteria to judge the quality of the assessment practices used by teachers and state and federal agencies to measure student achievement and the opportunity provided for students to learn science (NRC, 1996, 2001a, 2001b).

Standard A: Coordination With Intended Purposes

Assessments are consistent with the decisions they are intended to inform. What is expected in order to meet this standard?

- Assessments are deliberately designed.
- Assessments have explicitly stated purposes.
- The relationship between the data and the decisions should be clear.
- Assessment procedures need to be internally consistent.

Standard B: Measuring Student Achievement and Opportunity to Learn

Achievement and opportunity to learn science must both be assessed.

- Achievement data focus on the science content that is most important for students to learn.
- Opportunity-to-learn data focus on the most powerful indicators of the students' opportunity to learn.
- Equal attention must be given to the assessment of opportunity to learn and to the assessment of student achievement.

Standard C: Matching Technical Quality of Data With Consequences

The technical quality of the data collected is well matched to the decisions made and the actions taken on the basis of data interpretation.

- The feature that is claimed to be measured is actually measured.
- Assessment tasks are authentic.
- An individual student's performance is comparable on two or more tasks that purportedly measure the same aspect of student achievement.
- Students have adequate opportunity to demonstrate their achievements.
- Assessment tasks and methods of task presentation provide data that are sufficiently stable to lead to the same decisions if used at different times.

Standard D: Avoiding Bias

Assessment practices must be fair.

- Assessment tasks should be reviewed for the presence of stereotypes, for assumptions that reflect the perspectives or experiences of a particular group, for language that might be offensive to a particular group, and for other features that might distract students from the intended task.
- Large-scale assessments must use statistical techniques to identify potential bias that could contribute to differential performance.
- Assessment tasks must be appropriately modified to accommodate the needs of students with physical disabilities, learning disabilities, or limited English proficiency.
- Assessment tasks must be set in a variety of contexts and engage students with varied interests and experiences, and tasks should not assume the perspective or experience of a particular gender, racial, or ethnic group.

Standard E: Making Sound Inferences

The inferences made from assessments about student achievement and opportunity to learn must be sound.

- When making inferences from assessment data about student achievement and opportunity to learn science, explicit reference needs to be made to the assumptions on which the inferences are based.

The *National Science Education Standards* communicate the following expectations for assessment practices in science (Table 2.6).

Table 2.6   Changing Emphases of Assessment

Less Emphasis Placed On	More Emphasis Placed On
- Assessing what is easily measured - Assessing discrete knowledge - Assessing scientific knowledge - Assessing to learn what students do not know - End-of-term assessments by teachers - Development of external assessments by measurement experts alone	- Assessing what is most highly valued - Assessing rich, well-structured knowledge - Assessing scientific understanding and reasoning - Assessing to learn what students understand - Students engaged in ongoing assessment of their work and that of others - Teachers involved in the development of external assessments

Source: National Research Council. (1996). *National Science Education Standards.* Washington, DC: National Academy Press; and National Research Council. (2001a). *Classroom assessment and the national science education standards.* Washington, DC: National Academy Press.

IMPLEMENTING ASSESSMENT GUIDELINES

In this section, some considerations for how to implement assessment guidelines in a classroom will be addressed. The ways described are by no means exhaustive, but rather, they represent an attempt to draw more attention for consideration of potential applications. Teachers hold the best position to make assessment decisions and most of the responsibilities for implementing assessment standards in the context of teaching.

Assessment Reminders for Educators

Keep in mind the following six points:

1. A given assessment is only one piece of information used to provide evidence for meeting learning targets or objectives. Assessment should be considered a tool in the service of the learner rather than as a teacher-effectiveness measure.

2. The integration of assessment tasks into real-world problem-solving contexts should be a goal of assessment. Multiple assessments are recommended, and an emphasis should be placed on monitoring students' performances throughout the learning process and not just in the culminating assessment.

3. When looking at student outcome measures, questions that should be raised are those of when and where students have had the opportunity to experience and learn the information being assessed. A responsible teacher needs to be aware of and in control of as many factors as possible to reach a fair judgment when a given assessment is used.

4. Assessment standards are typically set by individuals other than students, but the teacher should invite students to participate in the establishment of some criteria for their assessments. Students should have an opportunity to contribute to what they learn and should learn how to assess their own learning.

5. Peer evaluation, group performance, and participation can be considered in determining a final grade.

6. In addition to normative comparisons, personal growth and ability levels should be taken into account in some ways when assessing students. For students to become engaged in learning, the opportunities to experience some level of success seem to be necessary in classroom instruction and assessment.

Embedding Assessment in Teaching Practice

While planning instruction and assessments, guidelines assist in moving ideas into practice. Some of the following strategies can help guide the planning process:

Communication of Expectations

- Teacher expectations for assessment and how the curriculum will be implemented should be clearly conveyed and discussed with students. For example, teachers should discuss with students the instructional goals and evidence of learning, the kinds of data or works to be collected, how student performance will be assessed, and when student work is due.
- Student participation in the planning of what is to be learned and assessed is essential to promoting self-regulation of learning. Real-world problem-solving activities, rather than textbook or memorization-oriented tasks, are probably necessary to have meaningful student involvement in learning.

Communication and Monitoring During the Learning Process

- Make certain the flow of curriculum is flexible enough to cope with the interaction in the classroom and retain the "how, what, and when" design initially indicated.
- Maintain a supportive learning atmosphere.
- Many assessment methods can be used to document student learning and modify the curriculum accordingly. For example, teacher observations of students could be used to monitor engagement.
- Students need time to internalize learning expectations, especially if they have not experienced more student-centered and open-ended learning environments.

Final Interpretation of Student Performance

- Validity and reliability of the assessments used to interpret and make decisions about student performance should always be scrutinized.
- As the teacher, be certain that claims made about student performance are based on both sound assessments and data interpretation.
- Teachers should use the information collected from student evaluations about what students perceive they are learning and doing. This feedback from students can inform the teacher in facilitating a student-centered classroom. Ideally, if students are learning in ways that are best for them, student performance should improve.
- Collect and review student work samples to inform your instruction. Identify both what students are doing well and where improvements are needed. Use this information to refine your work as a teacher.

THE INTERNET AS AN ASSESSMENT RESOURCE

The Internet has a wealth of information on assessment. The user must be a discerning practitioner in the selection of assessments or ideas to be used in the classroom, but many excellent sites exist. The 10 regional educational laboratories are excellent beginning points for a variety of assessment information.

Numerous resources can be accessed via the Regional Educational Laboratory Program (http://ies.ed.gov/ncee/edlabs) or each individual resource can be accessed to find their focus areas.

1. The Education Alliance at Brown University (http://www.lab.brown.edu)

2. Laboratory for Student Success (http://www.temple.edu/lss/index.htm)

3. Edvantia (http://www.edvantia.org)

4. Southeastern Regional Vision for Education (http://www.serve.org)

5. Southwest Educational Development Laboratory (http://www.sedl.org)

6. Learning Point Associates (http://www.learningpt.org)

7. Mid-central Regional Educational Laboratory (http://www.mcrel.org)

8. Northwest Regional Educational Laboratory (http://www.nwrel.org)

9. WestEd (Western Region) (http://www.wested.org)

10. Pacific Resources for Education and Learning (http://www.prel.org)

As mentioned earlier in this chapter, with the current emphasis placed on assessment, many states have developed Web sites that include samples and exemplars for assessments. The *National Science Education Standards* (NSES) are available online (http://books.nap.edu/books/0309053269/html/index .html). For evaluation and assessment information and resources, the Clearing House on Assessment and Evaluation Web site is helpful (http://ericae.net). The National Center for Research on Evaluation, Standards, and Student Testing (CRESST) (http://www.cse.ucla.edu) has numerous publications available on a range of assessment-related topics.

The National Council on Measurement in Education (NCME) has also set out the *Code of Professional Responsibilities in Educational Measurement* (NCME, 1995). This NCME code can be accessed online (http://www.natd.org/ Code_of_Professional_Responsibilities.html).

NCME also has developed a Code of Fair Testing Practices in Education (accessible at http://www.ncme.org/pubs/pdf/CodeofFairTestingPractices.pdf). These codes are informational for those involved in various assessment-related practices. Numerous educational measurement books and other resources from research centers and organizations also provide quality sources of assessment references ranging from theory to practice.

3

Evaluating
Teaching Practice

In this chapter, some possibilities for ways to evaluate teaching practice are described: action research, video recordings, journals, self-evaluations, various instruments to evaluate classroom practice, and more are included.

ACTION RESEARCH

Teachers are in a central position to determine the direction and success of educational reform, and with educational reform linked to teaching practice and professional growth, this requires commitment on the part of teachers. Teachers must begin or continue to inquire about their own practice, and action research can be a vehicle to do so. Action research, as Hopkins (1993) noted, combines a substantive act with a research procedure. It is the engagement in an inquiry combined with a personal attempt to both understand and improve or reform the situation.

Gay and Airaisian (2000) indicated that four beliefs underlie action research. First, action research provides an opportunity for teachers to engage in professional growth, and second, teachers who wish to improve their practice need data to do so. Third, in the process, teachers can use research findings to improve practice, and fourth, teachers can examine research findings and make applications in their own settings. Mills (2007) noted that good teachers have always examined the effects of their teaching on student learning. Action research supports critical reflection on practice (Mertler, 2006).

When teachers collect data to answer questions about effective teaching practices, they function as researchers engaged in inquiry. In education, many people categorize this as action research, which means teachers can systematically investigate ways by which they can improve their teaching and student learning. The *action* component is taken by the teacher in striving to analyze various aspects of teaching and learning. The term *research* indicates that a research plan has been structured and ultimately could be made available for public critique in the form of professional publications or presentations. Alternatively, the findings could be used to document personal professional growth for teacher evaluations and self-improvement plans. Engaging in classroom action research means that teachers develop study plans that involve collecting data related to students' achievement, opportunities to learn, and teaching practices.

When teachers conduct action research, many research questions can be posed and numerous strategies for data collection are possible. The teacher should work to refine and focus the research question to be addressed. If questions are too global, the action research process can become an insurmountable task. A highly focused question and a good research design to address or answer the question are imperative for success. A recommendation for anyone initiating action research is to keep the research manageable within the context of the daily classroom activities. Anyone who has engaged in research likely noted that in conducting research there is never a shortage of questions. Questions lead to other questions.

One method available to examine student learning in the classroom is video recording, where the camera can become the eyes for a teacher who wishes to record student actions and interactions for later review and critique. Teachers may miss elements of student learning because other classroom duties take priority. Although audio recordings could be used in a similar way, some advantages of video use that have been identified by teachers include the opportunities to view and critique components in the following list (Table 3.1).

Table 3.1   Possibilities for Informing Practice From a Visual Record

Student Checks	*Teacher Checks*
• Student attentiveness • Student self-paced work • Student on-task behaviors • Student-student interactions • Student performance using science processes (observation, classification, measuring, communication) • Student creativity (project work, adaptation of new ideas or processes, generation of alternative hypotheses) • Student group participation	• Teacher-student interactions • Teacher involvement in student performances • Teacher questioning strategies • Teacher levels of questions • Teacher communication of expectations • Teacher presentation and facilitation skills • Identification of assessment opportunities • Foci for discussion with students • Teacher self-evaluation and reflection

The video recording, along with other classroom artifacts such as photographs of classroom activities, journals, learning logs, and audio recordings, can become part of a teaching portfolio. The development of a set of written criteria to frame the use of information from video recordings is recommended. This can provide a focus for the documentation of events that can serve as a benchmark to which future data can be compared.

For teachers seeking or renewing National Board Certification, a video recording of classroom practice is required; more information on this can be found at The National Board for Professional Teaching Standards (NBPTS) (http://www.nbpts.org).

NBPTS certification is based on five core propositions (http://www.nbpts .org/the_standards/the_five_core_propositio):

1. Teachers are committed to students and their learning.

2. Teachers know the subjects they teach and how to teach those subjects to students.

3. Teachers are responsible for managing and monitoring student learning.

4. Teachers think systematically about their practice and learn from experience.

5. Teachers are members of learning communities.

With video recording and review of the recording, various components of practice can be identified for refinement. To improve practice, prioritizing and deciding what actions are to be taken can be supported with observation instruments that provide indicators and descriptors of practice. These indicators can be set out in the form of rubrics or checklists, and some instrumentation is available already in the literature or even online. A number of classroom learning environment questionnaires can also serve as tools to explore how students respond to teaching practices. For the teacher, this can become an action research project based on teaching practice.

Engaging in classroom research for self-improvement is likely not something to which students' parents and guardians will object, but the teacher may want to inform parents and guardians about the research through a letter that indicates intent and how the information will be used. Students are human subjects, and certain permission and approval may be required when conducting this kind of research—especially if done in conjunction with a university or college. Confidentiality of information should also always be considered. Administrators and parents are typically very supportive of this approach to improving professional practice.

Teacher journaling is another tool that can help teachers to reflect on which or how content pedagogical strategies have been implemented in the classroom. Teacher journal writing encourages thinking and can even serve the important function of integrating course content, self-knowledge, and practical experiences with teaching and learning situations. Journal entries can also provide evidence for a longitudinal professional growth profile. To be useful as a reflective tool, journaling needs to become a habit so that the events of reflection and levels of reflection can indicate patterns and changes. Journal entries could focus on (a) personal beliefs and knowledge about teaching and learning; (b) student responsiveness to the instructional strategies used in class; (c) classroom applications of information learned through literature, colleagues, and inservice workshops; and (d) personal reflections and feelings about the teaching and learning process.

A teacher journal related to constructivist teaching experiences framed on a Science-Technology-Society (STS) approach is shown in Figure 3.1 as an example. The STS unit focused on nutrition and the heart. A self-review of the entries can also provide insight on how reflective a teacher has been, and questions to frame the analysis might include, Is what I have written only a description of activities that occurred, or is what I have written really reflective? In what ways are the entries reflective?

Figure 3.1   Reflective Teaching Journal: An Example From Practice

Entry 1:	Our class is doing a nutrition/heart unit for our 20-day module. We discussed a food guide pyramid (how many servings are needed, etc.). They were given a copy of the pyramid to keep. We will be making a booklet of all our information throughout our unit. They seemed pretty excited about the topic. Hopefully it will stay that way!
Entry 2:	I gave the class a chance to come up with any ideas about nutrition that they would like to do or learn about. I was kind of disappointed because they didn't really come up with anything original or helpful. I have lots of ideas myself, so hopefully we'll do fine! They took a pretest on general nutrition/heart questions. The results weren't too bad, but I will be looking for lots of improvement and for them to become more knowledgeable.
Entry 3:	We went into more detail with the five food groups (the required servings, etc.). They got into groups and chose their favorite pizza. Then they came up with what food groups and servings were in it or on it. We all shared the results.
Entry 4:	The students are going to record what they eat for 3 days, starting today. They also filled out their own food pyramid to refer to. I'm finding I don't have as much time to get everything done like I'd like to in these class periods.
Entry 5:	We are keeping and comparing the students' 3-day eating record to the pyramid. Nearly everyone found some room to improve on at least one of the days! We started discussing physical fitness.
Entry 6:	We discussed different types of physical activities, and their favorites, and how it's good to do these for fitness. They got into groups and had a list of various activities. They marked boxes as to whether that activity falls under warm-up, strength, endurance, flexibility, or cooldown.
Entry 7:	We had a little more discussion about physical fitness. They also wrote on a paper what activities they did with their families, what they might add to their activities with their families, and what they could try.
Entry 8:	We are including self-esteem with nutrition and fitness. I gave them a chance to be specific about their strengths and accomplishments. They filled out a poster on "What I Like About Me," "What I'm Good At," and "Things That Make Me Proud." We shared a few things with the class.
Entry 9:	The students made a wellness pyramid to be tossed like dice. They would toss it and select a challenging question to ask or answer in order to review for a quiz on Monday.
Entry 10:	We didn't have school yesterday because of weather, so we had our quiz today on the food guide pyramid, and so on. I'm anxious to grade it tonight to see how they did. I just finished grading the quiz; I was very pleased with the results. Everyone improved on his or her score from the beginning pretest. I've got lots more I want to cover. I hope I have enough days to do it in!

Entry 11: I video recorded myself today during class. It went pretty well. I think the kids were more excited about it than I was! They divided up into groups of five and made two school lunch menus to be chosen during a week of school next month. They had to follow the requirement of having all five food groups in their menu. They were enjoying it!

Entry 12: They continued working on their menus. We got together and chose which ones to use and also discussed it with our cook. From February 7 through 11, we'll get to have these original menus.

Entry 13: We began talking about the heart (its purpose, parts, etc.). We watched a video on how the heart works and discussed it afterward. They had a fun sheet about the heart and what we talked about after that. Also, a tape on "Interviewing a Heart."

Entry 14: I video recorded a little bit today on the heart. We had some discussion and used posters and did a sheet for their booklets. I found out I use my hands a lot and could give more waiting time for the questions I ask my students.

Entry 15: I video recorded myself for my final part on tape. We discussed more on the heart and did a demonstration, which they got to participate in. We showed how a valve in your heart works by squeezing a demonstrator. We then used a siphon and "blood" and showed how your heart pumps blood, how it can pump faster when exercising, how it always goes one way, and so on. They had fun with it! I'm finding I have many more things to cover and not enough time! I know I will go over 20 days in this!

Entry 16: We practiced taking our pulses—resting and exercising. Some were surprised at the difference. I'm working on getting a speaker to come in and talk to the class. That has been difficult for me, to get a hold of someone—I'm still trying, though. The kids didn't know of anyone, so it was up to me! We studied for a quiz on the heart tomorrow. They know their information well!

Entry 17: We took our heart quiz. We also decided to put smoking into the heart/nutrition unit. We took a pretest to see what they knew. We also did a tar demonstration to see how it collects in their lungs (poured dark syrup into a measuring cup—they said when; they predicted how much collected in a person's lungs in a year).

Entry 18: We watched a smoking video. We did a Mystery Bag Activity. They had to guess what was in the bag by asking two students clue questions. They answered yes or no until they guessed what was in the bag by gathering information and narrowing possibilities—which is the same thing scientists and doctors have to do to figure out what causes heart disease, or to figure out how smoking affects your heart.

Entry 19: We took the smoking pretest again—they scored much, much better after all of our discussing and other activities. We related back to eating right and nutrition and talked about snacks (what's good for you, etc.). They divided up into groups and talked about healthy snacks.

Entry 20: The groups are coming up with two original snacks of their own. We will be making these into a recipe booklet and also trying some of them—sampling! (Yum!) They were also expected to mark off which food groups the ingredients came from. They enjoyed working on them.

Entry 21: We each had a partner today and were "Racing Raisins." (It dealt with carbon dioxide.) The raisins were placed in soda pop. After sinking, they watched how many times they got to the surface. (They could do what they wanted to help it get to the surface.) It was a challenge.

(Continued)

Figure 3.1 (Continued)

Entry 22: We were all Science "Fizz Whiz" people! We were in pairs and coming up with how soda is different from all other liquids (has carbon dioxide). We then did an activity where they put different things in soda (sugar, salt, sand, cotton swab, etc.) and watched what happened. They wrote or recorded what happened after each one. We discussed it afterwards. The kids thought it was great!

Entry 23: We collected menus from various restaurants. We compared them, seeing if they had all the food groups somewhere, what kinds of foods were available, and so on. They got into groups and talked about this and then reported to the class on the different ones. They learned which nutritional restaurants to go to!

Entry 24: Groups again! They decided which food group to use and are building a person using foods from that group. They will bring the ingredients needed in a few days. Also continued menu reporting.

Entry 25: Our speaker came today and talked about nutrition and the heart. She had had bypass surgery and knew lots about nutrition. The kids were interested and asked lots of questions. I wanted a dietitian but couldn't find one. Next month, a nurse will be coming, which should prove to be valuable also. She's planning on taking blood pressures, and so forth. We'll look forward to that.

Entry 26: The foods were brought for the person they'll make. They constructed them today and had fun. We had six different ones. They only used either the fruit or vegetable group. I thought we might have more of a variety. They turned out well. We will show them to the preschool tomorrow. We also worked on the skits.

Entry 27: The preschool loved the "food persons." Gave them some good examples for snacks. We also worked on our skits. They will be presented in a couple of weeks. The menus they made up were eaten at lunch all week this week. The kids were famous!

Entry 28: We played some nutrition games using what we learned. We reviewed other things for a miniquiz.

Entry 29: We had a miniquiz. They did wonderfully. We also played a game as a review. They then webbed and put up everything they'd learned. It was wonderful compared to when we'd started. I'm wrapping up my journal—we have a few loose ends to cover, but will finish up eventually. It's been fun; lots of work involved, but it was worth it. I'm ready to start another unit of STS.

ASSESSMENT OF CLASSROOM PRACTICE WITH VIDEO RECORDING SUPPORT

The use of video recordings to inform personal teaching perspectives aligns well with the construct of the teacher as a reflective practitioner. Video recordings of class lessons can provide feedback to inform teaching practice from self-evaluation of the videos. Peer review of video recordings can also be a very powerful tool for constructive feedback and can provide an opportunity for others to see examples of exemplary teaching. The forms that follow (Figures 3.2, 3.3, and 3.4) can be

Figure 3.2 Assessment of Classroom Interactions From Video-Recorded Sessions

Teacher Name: _____

Video Recording Context	Time Spent Dispensing Information	Time in Front of Classroom	Time With Individual Students or Groups	Use of Student Questions: Kinds and Levels of Questioning
Recording 1: Date, Context				
Recording 2: Date, Context				
Recording 3: Date, Context				
Recording 4: Date, Context				
Category Average: Summary Comments				

Comments/reminders for next steps:

Target for a focus on changing practice:

Figure 3.3 Finding Evidence of Effective Teaching Practices in Classroom Videos

---

Date and Lesson Context: _____

Main Lesson Concept(s): _____

1. Expectations for Student Learning   <u>Reminders to Myself</u>

Were expectations for what students   Yes   No
were to know and be able to do
communicated in student-friendly
language?

What evidence supports these observations?

2. Questioning Strategies

Were various levels of questioning used?   Yes   No
What levels of questions were asked?

Did the questions elicit student thinking   Yes   No
on various cognitive levels?

Was wait time practiced?   Yes   No

Were students engaged in asking   Yes   No
questions relevant to what they were
learning?

Did all students have the opportunity to   Yes   No
respond to questions?

How did I respond to student answers?

What kinds of feedback did I provide?

3. On-Task Behavior

Were the students on task?   Yes   No

What evidence supports these observations?

Did the room arrangement support on-   Yes   No
task behavior?

If students were not on task, what
changes could be made in this lesson
before future use?

What evidence supports these observations?

4. Use and Modeling of Instructional Strategies

Were appropriate instructional strategies used during the lesson?	Yes	No
Were appropriate instructional strategies modeled during the lesson?	Yes	No
Were multiple learning styles addressed during the lesson?	Yes	No
Did students have opportunities to discuss what they were learning?	Yes	No
Did students have the opportunities to write about what they were learning?	Yes	No

5. Student Assessment

Was assessment embedded in the lesson?	Yes	No

What kinds of assessments were used?

What kinds of feedback were provided for students?

Priorities for my teaching practice:

Figure 3.4   Self-Review of Classroom Teaching Video

Date and Lesson Context: _____

For each of the sections, A, B, and C, record the number of times the following were observed during each of two selected 10-minute periods.

A. Initiatory, informational	(Early 10 min.)		(Late 10 min.)	
Talking (discussion, lecture, or directions):				
Questions (initiatory):	Yes/No	Open-Ended	Yes/No	Open-Ended

Note to self:

Strengths:

Improvement target:

B. Responding	(Early 10 min.)		(Late 10 min.)	
	Yes/No	Open-Ended	Yes/No	Open-Ended
Teacher-centered (rejects or accepts student comments, confirms answer, repeats question or comment, clarifies or interprets, answers question directly):				
Student-centered (asks students to clarify and/or elaborate):				
Teacher-facilitated extensions (teacher discussion extending from student comment or question):				

Notes to self:

Strengths:

Improvement target:

C. Wait Time	(Early 10 min.)		(Late 10 min.)	
	Yes	No	Yes	No
(after the question) teacher ↔ student				
Wait Time 2 (after the first student response) student ↔ student				

Notes to self:

Strengths:

Improvement target:

D. On or Off Task	10 min.		15 min.		20 min.		25 min.		30 min.		35 min.		40 min.		45 min.	
On or Off Task	on	off	on	off	on	off	on	off	on	off	on	off	on	off	on	off

(During the observation, record the number of students on or off task at as many of the following intervals as possible.)

Notes to self:

Strengths:

Improvement target:

Source: Adapted from the SATIC coding form developed by Varrella, Kellerman, & Penick (1993). In R. E. Yager (Ed.), *Student Teaching Handbook.* Iowa City: University of Iowa, Science Education Center.

modified to conform to classroom needs. For example, if the teacher wished to focus on the kinds of questions being asked, he or she could provide descriptors and codes of question categories and then enter the code and tally in the matrix.

SOURCES OF INSTRUMENTATION FOR ASSESSING PRACTICE

Most school districts have evaluation instruments that are used to judge teaching performance, and this may be the context in which teachers begin. Using the district evaluation tools can help teachers provide better evidence of how criteria are met and interpreted. Furthermore, it can lead to selection of better instrumentation on which to be evaluated. Universities and colleges typically have protocols and instruments that they use in preservice programs, and these may be made available for use by contacting the education department of an area college or university.

Facilitating an inquiry-based classroom where students are engaged in doing science may require changes in teacher behaviors and teaching practices. Changes in practice likely take time to internalize and also require a focus on what to change and why the change has the potential to enhance student learning. The Reformed Teaching Observation Protocol (RTOP) developed by the Evaluation Facilitation Group (EFG) of the Arizona Collaborative for Excellence in Preparation of Teachers (ACEPT) is an observational instrument designed to measure *reformed teaching* of science or mathematics. Reformed teaching is teaching that enacts constructivist practices and inquiry-based approaches. The RTOP can be accessed online (http://www.science.utep.edu/gk12/rtopmanual.pdf), and a reference manual that describes procedures used to validate the instrument is provided. The instrument is designed for classroom observation, but the instrument could be used by a teacher for self-evaluation and self-study of practice. If lessons were video recorded, the RTOP could potentially be used by the teacher to examine personal practice.

The 25-item RTOP examines teaching on the basis of three factors: lesson design and implementation, content knowledge both propositional and procedural, and classroom culture focused on communicative interactions and student-teacher relationships. Even if school districts have evaluation instruments that are used for classroom observation and reporting, the RTOP may provide information that can guide teachers toward more student-centered instruction.

Horizon-Research, Inc. (HRI), the external evaluation contractor for the National Science Foundation (NSF)–funded Local Systemic Change (LSC) projects in science and mathematics education, developed a number of instruments for examining classroom practice. Interview protocols, observational instruments, and questionnaires are available at the HRI Web site (http://www.horizon-research.com/instruments). Before these classroom observation instruments were used to conduct classroom evaluations, evaluators attended training sessions on use of the instruments. With numerous evaluators rating multiple classrooms at multiple sites, this kind of training was conducted to increase interrater reliability. Video recordings of classrooms were viewed, and classroom practice was evaluated and discussed to assist evaluators in interpretation and use of the protocols.

A possible way to incorporate these observation protocols and questionnaires in practice might be to use them to examine one's personal practice. The questions could provide insights into the kinds of attributes that characterize standards-based teaching and inquiry-based teaching. Perusal of these instruments may help generate a direction to be taken in professional practice. A professional development study group could also collaborate in video recording and then use either the RTOP or HRI instruments to evaluate practice. These instruments could be used to help guide the focus for the evaluation of practice and for the examination of personal teaching practices. The concern about controlling for data comparisons across multiple school sites is not the issue that it is with research studies.

Reflective Practice

The Expert Science Teacher Educational Evaluation Model (ESTEEM), developed by Burry-Stock (1993), includes a number of instruments designed to evaluate various facets of constructivist practice. The Interstate New Teacher Assessment and Support Consortium (INTASC) standards, while targeting new teachers, also provide a framework for teaching practice that could be used to evaluate the classroom practice of veteran teachers. INTASC standards are online at various sites; to access them, use one of the search engines and type in INTASC. An online search for instruments that could be used in the classroom will likely locate other instruments, but it is a good idea to look for discussions related to the validity and reliability of the instrument.

SELF-ASSESSMENT OF CONSTRUCTIVIST PRACTICE

As Brooks and Brooks (1993) noted, it sounds like a simple proposition that we construct our own understandings of the world, but moving constructivism into teaching practice may take time and teacher awareness of indicators of constructivist practice. Figures 3.5, 3.6, and 3.7 may be useful in raising awareness in the move to a more constructivist classroom.

SCIENCE-AS-INQUIRY SURVEYS

As a teacher, your responses to the Science as Inquiry Survey (Figure 3.8) can provide evidence and documentation for your perceptions of your classroom practice. The student form of the survey (Figure 3.9) provides an opportunity for your students to input their perceptions. Comparisons can be made, and the data can be used to profile the science learning opportunities from both student and teacher perspectives (Figs. 3.10, 3.11, and 3.12). The NSES and *Benchmarks for Science Literacy* (American Association for the Advancement of Science [AAAS], 1993) were referents for the development of these survey forms.

Figure 3.5 Are You a Constructivist Teacher?

Teacher: _____ Date: _____

For statements 1 through 12, mark the letter that best indicates your perception of your teaching. Use the following categories:

4 = Often, 3 = Sometimes, 2 = Seldom, 1 = Never

1. I encourage and accept student autonomy and initiative.	4	3	2	1
2. I use raw data and primary sources along with manipulative, interactive, and physical materials.	4	3	2	1
3. When framing tasks, I use cognitive terminology, such as "classify," "predict," and "create."	4	3	2	1
4. I allow student responses to drive lessons, shift instructional strategies, and alter content.	4	3	2	1
5. I inquire about students' understanding of concepts before sharing my understandings of those concepts.	4	3	2	1
6. I encourage students to engage in dialogue, both with other students and with me.	4	3	2	1
7. I encourage student inquiry by asking thoughtful, open-ended questions, and I encourage students to ask questions of each other.	4	3	2	1
8. I seek elaboration of students' initial responses.	4	3	2	1
9. I engage students in experiences that might engender contradictions to their initial hypotheses and then encourage discussion.	4	3	2	1
10. I allow wait time after posing questions.	4	3	2	1
11. I provide time for students to construct relationships and create metaphors.	4	3	2	1
12. I nurture students' natural curiosity through frequent use of the learning cycle model.	4	3	2	1

Note: Complete this self-perception survey at the beginning of the year, at mid-year, and at the end of the year. This can provide a self-check and target areas that you may want to change. Providing examples or scenarios from practice to support perceptions could be a very strong evidential accompaniment to this self-perception survey.

Source: Developed from the work of Brooks, J. G., & Brooks, M. G. (1993). *In Search of Understanding: The Case for Constructivist Classrooms.* Alexandria, VA: Association for Supervision and Curriculum Development.

Figure 3.6   Student Instructional Preferences and Self-Motivation Survey

Name: _____   Date: _____

Class: _____

This survey asks you to give your preferences about how you like to learn science and what motivates you in science. Circle the letter that best matches your viewpoint. Use the following categories:

a = Strongly Agree, b = Agree, c = Disagree, d = Strongly Disagree

1. I prefer that my teacher tell me about science rather than to read a science book.	a	b	c	d
2. I like it when I have to explain the results of my own experiment.	a	b	c	d
3. Studying alone, I learn more than by studying in a small group.	a	b	c	d
4. In science classes, I would rather listen to the teacher than do other activities.	a	b	c	d
5. I like to do experiments, which help me to understand the science I have learned.	a	b	c	d
6. I like the teacher to explain rather than have to learn from books.	a	b	c	d
7. Taking tests helps me know if I have understood what I have learned in class.	a	b	c	d
8. I find it difficult to listen to the teacher for a long period of time.	a	b	c	d
9. I understand science concepts better if I have to explain them in my own words.	a	b	c	d
10. I like working in small groups in science.	a	b	c	d
11. I like the science teacher to decide how we learn science.	a	b	c	d
12. I learn more from doing experiments than by listening to the teacher's explanations.	a	b	c	d
13. I feel confused when I read several books about the same science idea.	a	b	c	d
14. I like to have my science teacher correct my homework.	a	b	c	d
15. I prefer to listen to the teacher rather than learn from doing experiments.	a	b	c	d
16. I like to find out something without the teacher telling me how to do it.	a	b	c	d
17. One of the best ways for me to understand science is to discuss it in class.	a	b	c	d

(Continued)

Figure 3.6 (Continued)

18. I would learn more if I could choose which science topics I studied.	a	b	c	d
19. I find it difficult to do science experiments without instructions from the teacher.	a	b	c	d
20. When I am interested in a scientific idea, I like to read more about it.	a	b	c	d
21. I would rather be tested by the teacher than anyone else.	a	b	c	d
22. I find it difficult to understand science without the teacher's explanations.	a	b	c	d
23. I would rather find out about a scientific idea on my own than have it explained by the teacher.	a	b	c	d
24. When working in small groups, my classmates share what they know with me.	a	b	c	d
25. I like the teacher to tell me what I have to do when doing an experiment.	a	b	c	d
26. The best science classes are those when we do experiments.	a	b	c	d
27. Taking notes is more useful for learning than reading textbooks.	a	b	c	d
28. My classmates know better than the teacher does whether or not I understand science.	a	b	c	d
29. By taking notes, I make sure that I study what the teacher wants me to learn.	a	b	c	d
30. Solving problems is one of the best ways for me to understand science.	a	b	c	d
31. I express my ideas more easily when I am working in a small group.	a	b	c	d
32. The teacher's answers to the questions asked in class by my classmates help me understand science.	a	b	c	d
33. I enjoy doing experiments.	a	b	c	d
34. I would rather use computers to learn science than to listen to the teacher.	a	b	c	d
35. Taking a test is not the only way of finding out if I have understood science.	a	b	c	d
36. I like to get good grades, even if I have to work hard to do so.	a	b	c	d
37. In science classes, if I do not understand something, I look it up in a book.	a	b	c	d

38. I get worried if I cannot solve a problem in science.	a	b	c	d
39. I would rather have friends than to be the best in the class.	a	b	c	d
40. I do not like other classmates to know if I get a poor grade.	a	b	c	d
41. I like learning about the latest discoveries and inventions in science.	a	b	c	d
42. I do not mind working hard in science class as long as I learn something.	a	b	c	d
43. I care what my classmates think of me.	a	b	c	d
44. I like to compete with others for the best grades.	a	b	c	d
45. In the science lab, I like to mix different chemicals to find out what happens.	a	b	c	d
46. I am ashamed when I get a low grade on a test.	a	b	c	d
47. When doing experiments, I prefer to work with my friends.	a	b	c	d
48. I like the teacher to tell the rest of the class when I get good grades.	a	b	c	d
49. I like to find out about new ideas in science.	a	b	c	d
50. I like the teacher to praise my efforts in science.	a	b	c	d
51. Having good friends is one of the most important things at school.	a	b	c	d
52. I try to lead in class discussions.	a	b	c	d
53. I am interested in many scientific ideas that are not taught at school.	a	b	c	d
54. I try to pay attention to what the teacher says so that I will not miss anything important.	a	b	c	d
55. I do not mind it when classmates copy my problems or work.	a	b	c	d
56. I like to get the best grade on tests.	a	b	c	d
57. I like to find out more information than what the teacher tells me in class.	a	b	c	d
58. I like it when the teacher gives detailed explanations.	a	b	c	d
59. I do not mind lending my books and notes to classmates.	a	b	c	d
60. I like to be one of the first to finish my class work.	a	b	c	d

(Continued)

Figure 3.6 (Continued)

61. I want to know more about many new science topics.	a	b	c	d
62. I like the teacher to check my homework every day.	a	b	c	d
63. I like my classmates to help me in class.	a	b	c	d
64. I am more interested in the grade I get than in the mistakes I have made.	a	b	c	d
65. I am interested in finding out the answers when solving scientific problems.	a	b	c	d
66. I try hard to please the teacher with my work.	a	b	c	d
67. I like working with friends when working in small groups.	a	b	c	d
68. In class discussions, I like to be able to present the best ideas.	a	b	c	d
69. I enjoy reading books about science.	a	b	c	d
70. I like to do my best when doing my science homework.	a	b	c	d
71. I like working in small groups.	a	b	c	d
72. I like to show others my answer when they do not know how to do the work.	a	b	c	d
73. I like to learn about new ideas in science.	a	b	c	d
74. I like homework because I learn more.	a	b	c	d
75. When working in small groups, I do not care with whom I work.	a	b	c	d

Figure 3.7 What Happens in My Science Classroom? Student Form

Questionnaire purpose: This questionnaire asks you to provide your opinions about your science classroom. This is not a test, and your answers will not affect your grade. There are no right or wrong answers. Your answers will help your teacher improve your science classes.

For each sentence, circle the letter in the column at the right that best describes you or your views.

Learning About the World	Almost Always	Often	Sometimes	Almost Never
In this class:				
1. I learn about the world outside of school.	a	b	c	d
2. My new learning starts with problems about the world outside of the school.	a	b	c	d
3. I learn how science can be a part of my out-of-school life.	a	b	c	d
In this class:				
4. I get a better understanding of the world outside of school.	a	b	c	d
5. I learn interesting things about the world outside of school.	a	b	c	d
6. What I learn has *nothing* to do with my out-of-school life.	a	b	c	d
Learning About Science	Almost Always	Often	Sometimes	Almost Never
In this class:				
7. I learn that science *cannot* provide perfect answers to problems.	a	b	c	d
8. I learn that science is influenced by people's values and opinions.	a	b	c	d
9. I learn that science has changed over time.	a	b	c	d
In this class:				
10. I learn about the different sciences used by people in other cultures.	a	b	c	d
11. I learn that modern science is different from the science of long ago.	a	b	c	d
12. I learn that science is *inventing* theories.	a	b	c	d
Learning to Express Myself	Almost Always	Often	Sometimes	Almost Never
In this class:				
13. It's OK for me to ask the teacher "Why do I have to learn this?"	a	b	c	d
14. It's OK for me to question the way I'm being taught.	a	b	c	d
15. It's OK for me to complain about activities that are confusing.	a	b	c	d

(Continued)

Figure 3.7 (Continued)

Learning to Express Myself	Almost Always	Often	Sometimes	Almost Never
In this class:				
16. It's OK for me to complain about anything that prevents me from learning.	a	b	c	d
17. It's OK for me to express my opinion.	a	b	c	d
18. It's OK for me to speak up for my rights.	a	b	c	d
Learning to Communicate	Almost Always	Often	Sometimes	Almost Never
In this class:				
25. I get the chance to talk to other students.	a	b	c	d
26. I talk with other students about how to solve problems.	a	b	c	d
27. I explain my ideas to other students.	a	b	c	d
In this class:				
28. I ask other students to explain their ideas.	a	b	c	d
29. Other students ask me to explain my ideas.	a	b	c	d
30. Other students explain their ideas to me.	a	b	c	d

Source: Adapted from the work of Taylor, P. C., Fraser, B. J., & White, L. R. (1994, March). *A classroom environment questionnaire for science educators interested in the constructivist reform of school science.* Paper presented at the annual meeting of the National Association for Research in Science Teaching, Anaheim, CA.

Figure 3.8 Science As Inquiry: Teacher Perceptions of Science Class

The questions on this survey relate to elements of teaching practice in your science classroom. For all sections, please circle the choice that matches your perception.

Section 1: Science Classroom

Use the following rating scale for Questions 1–34: 5 = Very Often, 4 = Often, 3 = Sometimes, 2 = Seldom, 1 = Never

In your science class, how often do you have your students do the following?	*Very Often*	*Often*	*Sometimes*	*Seldom*	*Never*
1. Work in groups or teams	5	4	3	2	1
2. Work in groups or teams when they do science activities	5	4	3	2	1
3. Work individually	5	4	3	2	1
4. Do activities and experiments in science class	5	4	3	2	1
5. Work alone to do science activities and experiments	5	4	3	2	1
6. Design their own activities and experiments	5	4	3	2	1
7. Try activities or experiments that they design themselves	5	4	3	2	1
8. Test a hypothesis or question in their activities or experiments	5	4	3	2	1
9. Control variables when they do lab activities or experiments	5	4	3	2	1
10. Ask questions and then investigate their own questions	5	4	3	2	1
11. Make predictions about what will happen before they do activities or experiments	5	4	3	2	1
12. Set up a data table when they do activities or experiments	5	4	3	2	1
13. Make observations when they do activities or experiments	5	4	3	2	1
14. Write down their observations from an experiment	5	4	3	2	1
15. Write about the experiments that they do in a notebook, log, or journal	5	4	3	2	1
16. Write down their own information from a science experiment	5	4	3	2	1
17. Graph numbers from their experiments	5	4	3	2	1
18. Discuss the results from their experiments	5	4	3	2	1

(Continued)

Figure 3.8 (Continued)

In your science class, how often do you have your students do the following?	Very Often	Often	Sometimes	Seldom	Never
19. Set up their own experiments or activities	5	4	3	2	1
20. Try experiments more than one time to check their results	5	4	3	2	1
21. Read about the research work that scientists do	5	4	3	2	1
22. Discuss the research work that scientists do	5	4	3	2	1
23. Discuss science articles from newspapers, magazines, or the Web	5	4	3	2	1
24. Go to the school library or media center to find science information	5	4	3	2	1
25. Watch and then discuss science DVDs or multimedia presentations	5	4	3	2	1
26. Have visitors come to class to talk about science	5	4	3	2	1
27. Go on field trips on campus (school grounds) that relate to what they do in science class	5	4	3	2	1
28. Go on field trips off campus that relate to what they do in science class	5	4	3	2	1
29. Student-led demonstrations and experiments for the class	5	4	3	2	1
30. If activities or experiments do not appear to work as predicted, we discuss reasons why.	5	4	3	2	1
Who decides what science lessons and activities are done in science class?					
31. I decide what the science lessons are about.	5	4	3	2	1
32. The students in the class decide what some of the science lessons are about.	5	4	3	2	1
33. I decide what science activities and experiments we do.	5	4	3	2	1
34. The students in the class decide about some of the science activities and experiments we do.	5	4	3	2	1

Section 2: Science Assignments

Rating scale for Questions 35–44: 5 = Very Often (3 to 5 times a week), 4 = Often (1 or 2 times a week), 3 = Sometimes (1 or 2 times a month), 2 = Seldom (1 or 2 times a year), 1 = Never (not done in science class)

In your science class, how often do you have your students do the following?	Very Often	Often	Sometimes	Seldom	Never
35. How often do you give science assignments?	5	4	3	2	1
36. Do students work in groups to complete some assignments?	5	4	3	2	1
What kinds of assignments do you use in your science class?					
37. Students answer questions at the end of a section or chapter in the science textbook.	5	4	3	2	1
38. Students write definitions of science words.	5	4	3	2	1
39. Students do science worksheets.	5	4	3	2	1
40. Students do concept maps, mind maps, or webs.	5	4	3	2	1
41. Students do assignments that require presentations.	5	4	3	2	1
42. Students do assignments that require the use of a variety of resources.	5	4	3	2	1
43. Students do assignments that require projects.	5	4	3	2	1
44. Students do assignments that require them to involve people from the community.	5	4	3	2	1

Section 3: Tests and Assessments

Rating scale for Questions 45–58: 4 = Frequently (for most tests or assessments), 3 = Sometimes (for some tests or assessments), 2 = Seldom (for very few tests or assessments), 1 = Never (not used in science class)

How often do you have the following on tests or for assessments?	Frequently	Sometimes	Seldom	Never
45. True-false questions	4	3	2	1
46. Multiple-choice questions	4	3	2	1
47. Matching questions	4	3	2	1
48. Fill-in-the-blank questions	4	3	2	1
49. Short-answer questions	4	3	2	1
50. Essay questions	4	3	2	1

(Continued)

Figure 3.8 (Continued)

How often do you have the following on tests or for assessments?	Frequently	Sometimes	Seldom	Never
51. Short-term products or projects (that take about 1 week to do)	4	3	2	1
52. Longer-term products or projects (that take more than 1 week to do)	4	3	2	1
53. Write reports about science	4	3	2	1
54. Maintain a science portfolio of their work	4	3	2	1
55. Make presentations about their work	4	3	2	1
56. Do concept maps, mind maps, or webs	4	3	2	1
Do you involve students in the following?				
57. Use student input for design of a scoring guide (rubric) for their work	4	3	2	1
58. Use student input for decisions on how some science work is graded	4	3	2	1

Section 4: Equipment and Materials Use

Rating scale for Questions 59–70: 5 = Very Often, 4 = Often, 3 = Sometimes, 2 = Seldom, 1 = Never

Do students use any of the following equipment or materials in science class?	Very Often	Often	Sometimes	Seldom	Never
59. Balances or scales	5	4	3	2	1
60. Thermometers	5	4	3	2	1
61. Microscopes	5	4	3	2	1
62. Magnifying lenses	5	4	3	2	1
63. Meters sticks or rulers	5	4	3	2	1
64. Timers or stopwatches	5	4	3	2	1
65. Computers for word processing	5	4	3	2	1
66. Computers with probes or science software	5	4	3	2	1
67. Computer simulations	5	4	3	2	1
68. Live animals or plants	5	4	3	2	1
69. Preserved animals or plants	5	4	3	2	1
70. Graduated cylinders to measure liquids	5	4	3	2	1

Source: Enger, S. (1997). *The Relationship Between Science Learning Opportunities and Ninth Grade Students' Performance on a Set of Open-Ended Questions.* Unpublished doctoral dissertation, The University of Iowa, Iowa City.

Figure 3.9 Science as Inquiry: Student Perceptions of Science Class

The questions on this survey relate to things that you may do in your science class when you are learning about science. For all sections, please circle the choice that matches your perception.

Section 1: Science Classroom

Use the following rating scale for Questions 1–34: 5 = Very Often, 4 = Often, 3 = Sometimes, 2 = Seldom, 1 = Never

In your science class, how often do you do the following?	Very Often	Often	Sometimes	Seldom	Never
1. Work in groups or teams	5	4	3	2	1
2. Work in groups or teams when you do science activities	5	4	3	2	1
3. Work by yourself	5	4	3	2	1
4. Do activities and experiments in science class	5	4	3	2	1
5. Work by yourself when you do science activities and experiments	5	4	3	2	1
6. Design your own activities and experiments	5	4	3	2	1
7. Try activities or experiments that you design yourselves	5	4	3	2	1
8. Test a hypothesis or question in your activities or experiments	5	4	3	2	1
9. Control variables when you do lab activities or experiments	5	4	3	2	1
10. Ask questions and then investigate your own questions	5	4	3	2	1
11. Make predictions about what will happen before you do activities or experiments	5	4	3	2	1
12. Set up a data table when you do activities or experiments	5	4	3	2	1
13. Make observations when you do activities or experiments	5	4	3	2	1
14. Write down your observations from an experiment	5	4	3	2	1
15. Write about the experiments that you do in a notebook, log, or journal	5	4	3	2	1

(Continued)

Figure 3.9 (Continued)

In your science class, how often do you do the following?	Very Often	Often	Sometimes	Seldom	Never
16. Write down your own information from a science experiment	5	4	3	2	1
17. Graph numbers from your experiments	5	4	3	2	1
18. Discuss the results from your experiments	5	4	3	2	1
19. Set up your own experiments or activities	5	4	3	2	1
20. Try experiments more than one time to check your results	5	4	3	2	1
21. Read about the research work that scientists do	5	4	3	2	1
22. Discuss the research work that scientists do	5	4	3	2	1
23. Discuss science articles from newspapers, magazines, or on the Web	5	4	3	2	1
24. Go to the school library or media center to find science information	5	4	3	2	1
25. Watch and then discuss science DVDs or multimedia presentations	5	4	3	2	1
26. Have visitors come to class to talk about science	5	4	3	2	1
27. Go on field trips on campus (school grounds) that relate to what you do in science class	5	4	3	2	1
28. Go on field trips off campus that relate to what you do in science class	5	4	3	2	1
29. Lead demonstrations and experiments for the class	5	4	3	2	1
30. If activities or experiments do not appear to work as predicted, we discuss reasons why.	5	4	3	2	1
Who decides what science lessons and activities are done in science class?					
31. The teacher decides what the science lessons are about.	5	4	3	2	1
32. The students in the class decide what some science lessons are about.	5	4	3	2	1

In your science class, how often do you do the following?	Very Often	Often	Sometimes	Seldom	Never
33. The teacher decides what science activities and experiments we do.	5	4	3	2	1
34. The students in the class decide about some of the science activities and experiments we do.	5	4	3	2	1

Section 2: Science Assignments

Rating scale for Questions 35–44: 5 = Very Often (3 to 5 times a week), 4 = Often (1 or 2 times a week), 3 = Sometimes (1 or 2 times a month), 2 = Seldom (1 or 2 times a year), 1 = Never (not done in science class)

In your science class, how often do you do the following?	Very Often	Often	Sometimes	Seldom	Never
35. How often do you have science assignments?	5	4	3	2	1
36. Do you work in groups to complete some assignments?	5	4	3	2	1
What kinds of assignments do you have in your science class?					
37. We answer questions at the end of a section or chapter in the science textbook.	5	4	3	2	1
38. We write definitions of science words.	5	4	3	2	1
39. We do science worksheets.	5	4	3	2	1
40. We do concept maps, mind maps, or webs.	5	4	3	2	1
41. We do assignments that require us to make presentations.	5	4	3	2	1
42. We do assignments that require us to use a variety of resources.	5	4	3	2	1
43. We do assignments that require us to complete projects.	5	4	3	2	1
44. We do assignments that require us to involve people from the community.	5	4	3	2	1

Figure 3.9 (Continued)

Section 3: Tests and Assessments

Rating scale for Questions 45–58: 4 = Frequently (for most tests or assessments), 3 = Sometimes (for some tests or assessments), 2 = Seldom (for very few tests or assessments), 1 = Never (not used in science class)

How often do you have the following on tests or for assessments?	Frequently	Sometime	Seldom	Never
45. True-false questions	4	3	2	1
46. Multiple-choice questions	4	3	2	1
47. Matching questions	4	3	2	1
48. Fill-in-the blank questions	4	3	2	1
49. Short-answer questions	4	3	2	1
50. Essay questions	4	3	2	1
51. Short-term products or projects (that take about 1 week to do)	4	3	2	1
52. Longer-term products or projects (that take more than 1 week to do)	4	3	2	1
53. Write reports about science	4	3	2	1
54. Keep a science portfolio of your work	4	3	2	1
55. Make presentations about your work	4	3	2	1
56. Do concept maps, mind maps, or webs	4	3	2	1
Do you do any of the following in your science class?				
57. Help the teacher design a scoring guide (rubric) for your work	4	3	2	1
58. Decide on how some science work is graded	4	3	2	1

Section 4: Equipment and Materials Use

Rating scale for Questions 59–70: 5 = Very Often, 4 = Often, 3 = Sometimes, 2 = Seldom, 1 = Never

Do students use any of the following equipment or materials in science class?	Very Often	Often	Sometimes	Seldom	Never
59. Balances or scales	5	4	3	2	1
60. Thermometers	5	4	3	2	1

Do students use any of the following equipment or materials in science class?	Very Often	Often	Sometimes	Seldom	Never
61. Microscopes	5	4	3	2	1
62. Magnifying lenses	5	4	3	2	1
63. Meters sticks or rulers	5	4	3	2	1
64. Timers or stopwatch	5	4	3	2	1
65. Computers for word processing	5	4	3	2	1
66. Computers with probes or science software	5	4	3	2	1
67. Computer simulations	5	4	3	2	1
68. Live animals or plants	5	4	3	2	1
69. Preserved animals or plants	5	4	3	2	1
70. Graduated cylinders or containers to measure liquids	5	4	3	2	1

Source: Enger, S. (1997). *The Relationship Between Science Learning Opportunities and Ninth Grade Students' Performance on a Set of Open-Ended Questions.* Unpublished doctoral dissertation, The University of Iowa, Iowa City.

Figure 3.10 Approaches to Teaching: Inventory for Self-Assessment

Directions: Think about a course or a class that you teach, or if you are a preservice teacher, envision teaching a course or class in your area of preparation. Please indicate the degree to which you agree or disagree with each statement below by circling the appropriate letters to the right of each statement.

SA = Strongly Agree, A = Agree, UN = Uncertain, D = Disagree, SD = Strongly Disagree

1. In my interactions with students in this class, I try to develop a <u>conversation</u> with them about the topics we are studying.	SA	A	UN	D	SD
2. I believe it is important to <u>present</u> a lot of facts to students so that they know what they have to learn to succeed in this class.	SA	A	UN	D	SD
3. I believe that the assessments in this class should be an <u>opportunity</u> for students to reveal their changed conceptual understanding of course content.	SA	A	UN	D	SD
4. I set aside some teaching time so that the <u>students can discuss</u> with each other any difficulties that they encounter in studying for this course.	SA	A	UN	D	SD
5. In this class, I concentrate on <u>covering the information</u> that might be available from a good textbook.	SA	A	UN	D	SD
6. I encourage students to <u>restructure their existing knowledge</u> (about course content) and find new ways of relating the concepts.	SA	A	UN	D	SD
7. In teaching sessions for this class, I use challenging or ill-defined examples to <u>generate debate</u>.	SA	A	UN	D	SD
8. I think an important reason for the class sessions is to <u>make sure students have a good set of notes to use for studying to do well on tests</u>.	SA	A	UN	D	SD
9. In this class, I provide the students with only the <u>information they will need to pass</u> the formal assessments.	SA	A	UN	D	SD
10. I believe that I should <u>know the answers to any questions</u> that students ask during class sessions.	SA	A	UN	D	SD
11. In this class, I provide opportunities for students to <u>discuss their changing understanding, views, and opinions regarding the class content</u>.	SA	A	UN	D	SD
12. I believe that it is better for students to <u>generate their own notes</u> rather than merely copying mine from PowerPoint or teacher-provided handouts.	SA	A	UN	D	SD

Based on how you responded to this inventory, write a description that uses adjectives that illustrate your approach to teaching.

Source: Adapted from lead teachers in the Iowa Chautauqua Program, University of Iowa, Iowa City, Iowa, 2008.

Figure 3.11   Student Perceptions of Teaching: Questions for Students

Please indicate the degree to which you agree or disagree with each statement below by circling the appropriate letters to the right of each statement.

SA = Strongly Agree, A = Agree, UN = Uncertain, D = Disagree, SD = Strongly Disagree

1. In the teacher's interactions with students in this class, he/she tries to develop a <u>conversation</u> with us about the topics we are studying.	SA	A	UN	D	SD
2. The teacher feels it is important to <u>present</u> a lot of facts and details to us so that we know what we have to learn to succeed in this class.	SA	A	UN	D	SD
3. The teacher feels assessments (exams and assignments) in this course should be an <u>opportunity</u> for us to reveal our changed conceptual understanding in this class.	SA	A	UN	D	SD
4. The teacher sets aside some teaching time in most class sessions so that <u>we can discuss</u> with each other the difficulties that we have studying for this class.	SA	A	UN	D	SD
5. In this class, the teacher concentrates on <u>covering the information</u> that might otherwise be available from a good textbook.	SA	A	UN	D	SD
6. The teacher encourages us to restructure our <u>existing knowledge</u> (about the class material) in terms of the new ways of thinking about the content in this class.	SA	A	UN	D	SD
7. In teaching sessions for this class, the teacher uses difficult or undefined examples to <u>provoke debate</u>.	SA	A	UN	D	SD
8. The teacher thinks an important reason for class time in this course is to <u>give us a good set of notes to study what we need to know</u>.	SA	A	UN	D	SD
9. In this class, the teacher provides us with only the <u>information we will need to pass</u> the exams.	SA	A	UN	D	SD
10. The teacher feels that he/she should <u>know the answers to any questions</u> that we may ask him/her during class sessions.	SA	A	UN	D	SD
11. The teacher makes opportunities available for us in this course to <u>discuss our changing understanding, views, and opinions regarding the course material</u>.	SA	A	UN	D	SD
12. The teacher feels that it is better for us in this subject to <u>write and develop our own notes</u> rather than merely copying from PowerPoint or teacher-provided handouts.	SA	A	UN	D	SD

Source: Adapted from lead teachers in the Iowa Chautauqua Program, University of Iowa, Iowa City, Iowa, 2008.

Figure 3.12 Measuring Student Perceptions of Science

Please indicate the degree to which you agree or disagree with each statement below by circling the appropriate letters to the right of each statement. The abbreviations used are:

A = Always, F = Frequently, ST = Sometimes, S = Seldom, N = Never

About Your Science Teacher

1. Teacher asks questions.	A	F	ST	S	N
2. Teacher likes you to question.	A	F	ST	S	N
3. Teacher allows you to question.	A	F	ST	S	N
4. Does your teacher really like science?	A	F	ST	S	N
5. Does your teacher make science exciting?	A	F	ST	S	N
6. Does your teacher know a lot about science?	A	F	ST	S	N
7. Does your teacher admit not knowing answers to your questions?	A	F	ST	S	N

About Your Science Classes

8. Science classes are fun.	A	F	ST	S	N
9. Science classes are interesting.	A	F	ST	S	N
10. Science classes are exciting.	A	F	ST	S	N
11. Science classes are boring.	A	F	ST	S	N
12. Science classes are uncomfortable.	A	F	ST	S	N
13. Science classes make you feel successful.	A	F	ST	S	N
14. Science classes make you feel curious.	A	F	ST	S	N

Usefulness of Studying Science

15. Things learned are useful outside of school.	A	F	ST	S	N
16. Things learned will be helpful in the future.	A	F	ST	S	N
17. Science you study is generally useful to you.	A	F	ST	S	N
18. Science prepares you to deal with current social issues/problems.	A	F	ST	S	N
19. Science classes help you improve your health.	A	F	ST	S	N
20. Science helps in dealing with your curiosity.	A	F	ST	S	N
21. Science helps in making decisions.	A	F	ST	S	N
22. Science classes provide examples of what scientists and engineers do.	A	F	ST	S	N
23. Science classes provide examples of related science careers.	A	F	ST	S	N
24. Science classes focus on both technology and science.	A	F	ST	S	N

Being a Scientist

25. Would be fun.	A	F	ST	S	N
26. Would make me rich.	A	F	ST	S	N
27. Would be a lot of work.	A	F	ST	S	N
28. Would be boring.	A	F	ST	S	N
29. Would make me feel important.	A	F	ST	S	N
30. Would make me lonely.	A	F	ST	S	N

For items 31–38, circle *a* for the subjects you like best, and circle *b* for the subjects you do not like. Last, circle the name of your favorite school subject.

31.	Foreign Language	a	b		35.	Physical Education	a	b
32.	Language Arts	a	b		36.	Science	a	b
33.	Music	a	b		37.	Reading	a	b
34.	Mathematics	a	b		38.	Social Studies	a	b

For items 39–42, circle the letter for the item that is most accurate for you.

39.	Gender	a. female	b. male	
40.	My science grades	a. B or better	b. C	c. D or lower
41.	My family's socioeconomic level	a. high	b. middle	c. low
42.	My plans for college	a. will go	b. may go	c. will not go

Source: Adapted from National Assessment of Educational Progress. (1978). *The Third Assessment of Science* (1976–1977). Denver, CO: Author.

4

Rubrics and Scoring Guides

The design and classroom use of rubrics and scoring guides have been embraced by teachers for communicating expectations and assessing student work. Some ideas for rubric design and use are addressed in this chapter.

RUBRICS AND SCORING GUIDES: WHAT ARE THEY?

"Do you have a rubric for that?" In today's classrooms, elementary through university, rubrics are expected as a component of work—especially work that involves projects, products, and performances. The words *scoring guide* are often used interchangeably with the word *rubric*. Rubrics or scoring guides can be found on the Web, and tools and templates are available for adapting or generating rubrics to assess various kinds of student work. Also, software tools like RubiStar are available from Advanced Learning Technologies (ALTEC) at the University of Kansas Center for Research on Learning (http://www.altec.org).

As Arter and Chappuis (2006) noted, the use of rubrics to communicate expectations to students has influenced the increased emphases on formative assessment in the classroom. They note that the world is awash in rubrics! Even if ready-made rubrics are available, developing and designing rubrics can be informative for both students and teachers. In the rubric development process, the clarification of criteria for a product or project communicates both expectations for the work and attributes of the final product. Student involvement in rubric development can assist students in understanding criteria and various levels of performance.

Historically, *rubric*, from the Latin for red, referred to practices used to mark important passages in a liturgy. Education has used the word *rubric* to denote an established form or method, and rubrics or scoring guides can provide forms and methods for communicating criteria for activities, events, conceptual development, or goals for learners. Rubrics provide a framework or structure on which student work can be developed and assessed, and rubrics and scoring guides can be designed and adapted for a wide range of student work.

With the current emphases on accountability and assessment, having evidence of preparing students for success is probably more important than ever; well-planned instruction and assessments supported with scoring guides can provide evidence that we can identify the kinds and levels of performances we expect as educators. Also, the process of developing scoring guides can be participatory and include student input. Criteria for student outcomes have been viewed many times as end-of-the-year goals are met, usually charting growth and change one section, unit, or chapter at a time. Viewed as end results, student progress toward these outcomes may not be formatively assessed. Criteria should allow teachers to develop end-of-unit assessments that indicate how each student is progressing toward the goal of understanding district and other desired learning outcomes. Current understanding and reforms in science education depict learning as an active process in which the students learn individually, collectively, competitively, and collaboratively. Each of the aforementioned characteristics is present in the successful classroom, yet they are often not assessed because of district requirements for standardized testing, time constraints, or personal beliefs about teaching and testing.

To assess the objectives represented by the type of active learning desired for students, student understanding should be assessed prior to instruction, during the instructional phase, and following the instructional sequence. Assessment should be viewed as an ongoing component of the classroom. Although many argue that there is little time to do all of this assessment, a counterargument is that each student is an individual and no two students could possibly know exactly the same material or have the same mental constructs for any subject. For students to develop understanding, time must be focused on both formative and summative assessments, and the information collected should be used to inform instruction. The results of these assessments should then be used to develop the criteria each student will address prior to completing the instructional sequence. This is where rubrics or scoring guides can be used to focus thinking and learning for both the teacher and the student. If students become a part of the development process in designing rubrics, this can help them in understanding how to identify important and relevant criteria in their work.

Rubrics are generally categorized as either analytic or holistic, and the type of work product that is being assessed typically determines which kind of rubric is the best fit. If the performance can be separated in components or dimensions that can be judged independently, then an analytic rubric is likely the most appropriate (Arter & McTighe, 2001). As an example, writing performance can be examined on six traits—ideas, organization, voice, word choice, sentence fluency, and conventions—and an analytic rubric would be appropriate for its assessment. In science, an analytic rubric could be used with laboratory reports and data representation. When the work can be quantified in terms of elements present, the analytic approach may be the best fit. If the attributes and qualities of the work being assessed are more global and less easily quantified, then a holistic approach is usually the best choice. Stiggins (2008) noted that accomplished teachers are connoisseurs of good performance and can recognize and describe outstanding performances. This thinking captures the essence of the holistic approach.

This chapter includes examples of rubrics that have been developed and used by teachers; teachers and their students have created these for their specific needs. The sample of rubrics includes some that are related to biology, to concept mapping, and to physical science topics. The opportunities are limitless for adapting these to many learning situations. A selection of examples of not very good rubrics is also included at the end of the chapter.

RUBRIC CONSTRUCTION: A TEACHER'S PERSPECTIVE

A scoring rubric or grid is an essential part of an assessment model. The model conveys being up front with the students in terms of what is expected of them and at what level or standard the work should be completed. The rubric is based on expected outcomes, and the instructor asks him- or herself what he or she expects the student to be able to do after having completed the assigned work. These expectations become the outcomes that guide student products or performances. The expected outcomes become strands on the scoring rubric. Each of the strands indicates to the student the standards for exemplary work, acceptable work, and work that must be revised. The teacher might ask, "What is one outcome I might expect from a student writing a paper on a scientific investigation?" Perhaps one of the outcomes would be that the student made few scientific errors. As a result, the strand for that outcome might appear as follows:

3	2	1
No major errors in scientific investigation; high level of accurate information	Some scientific errors that distract the reader from the major significance of the information; some information inaccurate or omitted	Enough scientific errors to render the essay ineffective for useful information; important information highly inaccurate or omitted

The inclusion of one desired outcome in the descriptors may initially make the rubric easier for the instructor and students to use. The rubric is meant to communicate to the student that the accuracy of the investigation is of prime importance. The teacher would continue to build the rubric based on the desired outcomes for this particular assignment. Students could also be involved in building the rubric. The rubric could include other strands, such as insights the student was able to form based on completed research, the quality of the research materials used, and the ability to support a conclusion based on clear evidence. Each idea or outcome would create one strand of the rubric. The descriptions of the standards should be parallel across the score points. A strand on resource use for this rubric follows:

3	2	1
Indication of multiple use of resources; good use of research to support thesis	Research goes beyond basic text; indication of use of alternative viewpoints; some additional information would enhance support of writer's thesis	Little or no evidence of research beyond the basic textbook

It is important to include only one desired outcome in the descriptors for the standards in each strand. Otherwise, it becomes difficult to decide how to award points if the student achieves one part of a multiple strand but not all parts. If the strand addresses only one idea, then the determination about achieving the standard is clearer. Setting out the rubric or scoring guide in advance provides a communication guide for both the instructor and the students.

The instructor should be certain that all outcomes to be assessed are addressed in the rubric because it would be unfair to assess students on criteria not communicated by the rubric. If one reason for initially giving students the rubric is to demystify the way in which they will be assessed, it would be unfair to change the rubric without informing them. If a rubric has been prepared that does not address an outcome that is later thought to be desirable, then the rubric should be rewritten if the same assignment is to be given later. Rubric or scoring guide development tends to become an iterative process as descriptors are refined with repeated use of the rubric.

Rubrics may appear to be subjective because phrases like "no major errors" rather than "fewer than 10 errors" are used. If a rubric has standards that indicate exemplary work as providing four factors that contribute to global warming, the student is expected to equate quantity with quality. He or she may give four factors, but are they meaningful factors? Are there more than four important factors? Is the student more interested in finding four factors than in concentrating on their relative importance? Perhaps it is better to allow the student to decide for himself or herself how many important causes can be successfully supported as relevant. This allows the student to place more emphasis on quality of response rather than quantity.

Rubrics also serve the purpose of making the evaluation of papers and projects more focused. Students can be clearer about teacher requirements and expectations, which in turn can lead them to write better papers and produce better work. Teachers themselves can also be clearer about expectations. Therefore, reading the paper or assessing a project with the clear intent of evaluating already established criteria is likely to result in a fairer assessment of student work. No longer does the teacher assess work to discover the standards; the standards for performance have been established at the outset. Students can also submit a self-assessment of their work based on the rubric.

In summary, a rubric can serve students by

- clarifying teacher expectations for the students;
- setting criteria and standards for performance in advance;
- involving students in the process;
- allowing the student to explain what he or she knows about a subject, rather than determine the right answers as identified by the teacher;
- providing the student with a clearer picture of how to organize a paper or project;
- making the evaluation of student materials more focused and objective; and
- providing a self-assessment opportunity for the students.

RUBRIC OR SCORING GUIDE DESIGN

Before beginning the design of a rubric or scoring guide, consider the assessment design. *Purpose* is the key that should guide the development of any assessment. What is the purpose of the assessment? Does assessment align with the learning experiences? What are the intended outcomes that can be expected from students as a result of their learning experiences? What are the measurable student outcomes that can realistically be expected? The use of a rubric or scoring guide can be a mechanism to communicate assessment intentions and to provide feedback to the students about their performances. This assessment information is often combined in some way with other information for grading and reporting purposes.

Rubric design and development can be accomplished in multiple ways. A general framework can be set up, and then the general framework can be customized for specific assessments. The reverse strategy can also be used, with a specific rubric designed for each assessment that could then be generalized. The general template approach may work well since the overarching assessment framework is in place. The works of Nitko (1996), Nitko and Brookhart (2007), and Marzano, Pickering, and McTighe (1993) are three useful references for the design process; with the current emphasis on assessment, many other excellent sources are available in print or on the Web. Numerous states either have or are developing assessment frameworks, and both assessment samples and scoring guides are online at state department of education Web sites.

A general rubric grid could be organized like the example in Table 4.1. Three categories may suffice to assess the performance; 5, 3, and 1 are often used to head the columns. A "4" column and "2" column can also be inserted. A set of exemplars for each category can be collected over time and used to communicate to students what attributes exemplify each performance level. Constructive feedback is recommended if student work falls in a "not acceptable" category, and where feasible, the student can be provided an opportunity to resubmit the work. Equally important is the feedback to those whose work meets the "best" level so that the attributes can be met in future work.

Table 4.1 Sample Grid for a Rubric

5 (Best)	3 (Good)	1 (Fair—Improvement Needed)
This block usually addresses what the teacher and students consider to be the best possible work that could be done by a student or group. All criteria are met.	Work that meets this level of success is good but does not meet all of the criteria set by the teacher and students. Some errors or flaws are present in the work.	Criteria in this block indicate that the work completed clearly shows that it could be improved on. This area addresses the kind of work, attitude, or achievement that is considered to be unacceptable by the teacher and students.

It is important for students to have some level of input in setting the criteria for their work because it empowers them in their assessment. One strategy that can be used to gain students' input is to provide them with a draft rubric and ask them to provide or help write descriptors and explanations for each performance level. After this process, students should have a better understanding of how to meet the expected outcomes.

EXAMPLES OF TASKS AND RUBRICS

Examples of a variety of tasks and rubrics are presented in the following pages.

Rubric for Interdisciplinary or Specified Domain Assessment

Table 4.2 shows a rubric that includes elements that address expectations for the generation of a model, a presentation, or both. In rubric design, consider the number of categories to be established because as more categories are established, more score points are needed, and this sometimes makes it more difficult to establish descriptors of performance. Although the rubric template is set out with five categories, it could be collapsed to three with score points of 5, 3, and 1, which might make it easier to assess performance. When a product or project is complex, trying to isolate individual pieces can become very challenging since you are assessing work with varied levels of complexity.

The development of criteria and the resulting rubric are presented in Table 4.3. This style of rubric was designed for the assessment of student work from a biology class that implemented a Science-Technology-Society (STS) approach.

Table 4.2 Rubric for Interdisciplinary or Specified Domain Assessment

Science Domain	5	4	3	2	1
Scientific thought (creativity, concepts, world view)	Original and creative design; strong evidence of understanding the scientific principles of _____ (insert topic or concepts here)	Attempt at an original design; expresses some creativity and provides some evidence of understanding the scientific principles	Model built from kit or plan; less creativity noted than with a more original design; some evidence of understanding of scientific principles	Model built from kit or plan; creative elements not apparent; limited evidence of understanding of scientific principles	Model built from kit or plan; lack of creativity; little or no evidence of understanding of scientific principles
Presentation (language arts, speech, conceptual understanding)	Clear and concise; effective use of scientific terms; complete understanding of topic for this age group; able to extrapolate	Well organized; good use of scientific terms; good understanding of topic; a few errors present	Presentation acceptable; adequate use of scientific terms; acceptable understanding of topic; more errors present	Presentation lacks clarity and organization; little inclusion of scientific terms and vocabulary; limited understanding of topic; numerous errors	No presentation developed; no intent to present
Exhibit (attitude, world view, use of technology, art)	Model completed to specifications; excellent workmanship; materials used in exemplary fashion; information is self-explanatory	Model completed to specifications; good workmanship; effective use of materials; information clear and logical	Model completed; adequate workmanship; materials used appropriately; acceptable information	Model partially completed; some problems with workmanship; materials could be used more effectively; some information lacking	Model not completed; poor workmanship; poor use of materials

Source: Adapted from Mason City City Schools. Mason City, Iowa.

Biology Assessment Task

A Sample Performance Task

The U.S. Surgeon General has selected you and your colleagues to serve on a committee that is responsible for raising public awareness of an environmental issue that has a potential impact on human health. The U.S. Surgeon General would like you to identify an issue and develop a position statement and suggestions for addressing the issue. To assist you with the development of your position statement and recommendations, the following guidelines should help focus your task:

- Identify an environmental issue that has the potential to have an impact on human life.
- Prepare a position paper that addresses this issue and suggest ways to address this issue.
- Research the issue your group has chosen.
- Describe how this issue affects the human body systems.
- Address potential problems if this issue is not addressed or resolved.
- Develop a plan to educate others about your findings.

Table 4.3 Performance Task Rubric

Criteria	10 (Set your own exemplary standards)	8	6	4
Issue identified and group position stated		Significance of issue clearly presented, with detail	Significance of issue clearly presented but more detail needed	Issue stated but little said about the significance
Linkage of environmental effects on body systems		Two effects discussed per system; three additional effects identified and discussed	Two effects discussed per system; two additional effects identified and discussed	Two effects discussed per system; two additional effects identified and discussed
Evidence and support		Extensive support data included (What is extensive? Students can discuss this.)	Adequate support data included (What is adequate? Students can discuss this.)	Minimal support data included (What is minimal? Students can discuss this.)
Body systems				
Reproductive				

Criteria	10 (Set your own exemplary standards)	8	6	4
Respiratory				
Circulatory				
Nervous				
Other				
Plan to educate others		Complete education component for identified systems	Adequate education component for identified systems	Minimal education component for identified systems
		Creativity used to enhance means of education	Some creative endeavor attempted to enhance means of education	Minimum creative effort
Potential impact on life		Two examples of impact potential summarized per system; three additional examples included	Two examples of impact potential summarized per system; two additional examples included	Two examples of impact potential summarized per system; one additional example included
References (depth of research), reference page		Five different types of references; reference page included	Four different types of references; reference page included	Three different types of references; reference page included

Source: Adapted from Mason City Schools, Mason City, Iowa.

Engaging students in the process of generating a rubric—from establishing criteria to developing various performance levels—is very productive. Students can identify criteria categories and then help differentiate performance levels. Rubrics often require wordsmithing and fine-tuning. What level of grain size is necessary to communicate and assess the performance? Does the rubric focus on very fine-grained details, or does the rubric focus on the bigger picture? Instead of using numbers to communicate the various performance levels, words can be used. If discussions are conducted with students about projects or products, these can really assist students in internalizing expectations.

The rubrics that follow are samples of teacher-developed rubrics, and these are likely to need modification to address the uniqueness of each classroom (Tables 4.4–4.10). The score points and weighting for each performance level can be added at the discretion of the teacher and students.

Table 4.4 The Application Domain: A Proposal for Social Action

Distinguished Performance	Proficient Performance	Intermediate Performance
Proposal has realistic and practical applications	Proposal has merit but is not realistic and practical	Proposal not yet realistic
Proposal supported by data (case studies, example, statistics, etc.); misconceptions cleared up	Proposal weakened due to lack of supporting data, or a few misconceptions evident, or both	Proposal not supported by data
Resources accurately documented in given format	Resources documented—format not followed	Resources not yet accurately documented
Displays an understanding of the ethical question and comes up with a believable presentation of both sides	Both sides of the proposal presented	Both sides of the proposal not yet evident
Presents a variety of choices for possible ideas or solutions	Possible ideas or solutions not fully explained	Has not yet presented other possible solutions
Explains the economic impact of the proposal	Presents the economic impact of the proposal	Economic impact not yet presented
Explains how this proposal may affect future generations	Presents how this proposal may affect future generations	Has not yet presented how proposal affects future generations

Note: Realistic = persuasive, well grounded, and documented; practical = acceptable by society and has potential for implementation.

Source: Adapted from Mason City Schools, Mason City, Iowa.

Table 4.5 Rubric for Use in a Scientific Investigation

Distinguished Performance	Experienced Performance	Average Performance	Novice Performance
Student can apply all the skills needed to investigate a self-selected issue or problem	Student can apply appropriate skills needed to investigate a self-selected issue or problem	Student can apply some investigative skills to a problem identified by another person	Student has difficulty identifying skills needed to study a provided problem
Student can recognize the phenomenon being investigated and develop an explanation for it	Student can identify the phenomenon being searched for and understands its nature	Student has difficulty identifying phenomenon being searched for and cannot explain the findings	Student cannot identify phenomenon being researched and is unable to explain why there is difficulty

Distinguished Performance	Experienced Performance	Average Performance	Novice Performance
Student can report findings to fellow classmates and to other interested parties	Student can report findings to fellow classmates and make findings available to others	Student can develop a report to give to fellow classmates but has difficulty communicating to others	Student cannot report to fellow students or others the reasons for lack of identification of phenomenon
Student can generalize findings to new and unique situations, thereby using the findings as a tool for investigation	Student can generalize findings to other similar situations and makes appropriate connections	Student can compare findings to other similar situations but has difficulty making connections	Student is unable to recognize similar situations and cannot make connections

Source: Adapted from a rubric by Lawrence Kellerman for Mason City Schools, Mason City, Iowa.

The rubric shown in Table 4.6 was designed by a teacher to assess a student's role in a group presentation. Expectations for references and presentation format should be communicated to students in advance of the presentation.

Table 4.6   Student Role in Group Presentation

3	2	1
Explained and demonstrated very thoroughly his or her role in the presentation	Somewhat explained his or her role in the presentation; role not clearly understood	Said very little if anything about role he or she played in the presentation
Responsibility for his or her part of project was very obvious	Evidence of some time and thought put into part	Minimum of work was done—just enough to meet requirements
Shared ideas and feelings willingly in presentation	Some sharing occurred	Very little, if any, sharing
Background research very evident—four or more sources mentioned	Some background research evident—two or three sources	Very little evidence of research sources—none to one source
Visuals used were pertinent and easily read	Visuals used but not easy to understand	Used no media in presentation
Comments for student:		

A science project developed around an issue was the context for the rubric in Table 4.7. In the last category, space could be left so that the actual problems with the project could be written in and feedback could be provided to the student. Also, in this rubric, content has the same weight as other components; if content is of major importance, this area should perhaps receive more weight in the assessment. (*Note:* While writing mechanics are important, awarding a separate score for grammar and mechanics may be an alternative, and this is recommended by the assessment developers.) Students can also use the rubric to self-assess and provide a reason why they have assigned the point values to their work.

Table 4.7 Science Issue Project Rubric

3	2	1
Sources • Lists more than three sources • Includes title, page, date • Includes author	Sources • Lists three sources • Includes title, page, date • Includes author	Sources • Lists two sources • Includes title, page, date
Report title • Relates to main topic • Capitalized correctly • Captures reader's interest	Report title • Relates to main topic • Capitalized correctly	Report title • Relates to main topic
Content • Many related topics • Informative • Well organized	Content • Relates to topic • Informative	Content • Relates to topic
Paragraphs • Complete sentences • Correct spelling • Correct punctuation • Correct capitalization • Neatly done	Paragraphs • Complete sentences • Correct spelling • Correct punctuation • Correct capitalization • Neatly done	Paragraphs • Complete sentences
Illustrations • More than one—titled • Labeled correctly • Neat appearance	Illustrations • Titled • Labeled correctly • Neat appearance	Illustrations • Titled • Labeled
Activities • Relates to issue question • Informative • Written conclusion	Activities • Relates to issue question • Informative • Written conclusion	Activities • Relates to issue question
Cover • Title • Name • Date • Illustration	Cover • Title • Name • Date	Cover • Title • Name

Note: All reports will be revised until an A, B, or C is earned. No grades D or F will be awarded.

Table 4.8 Peer Teaching: Group Rubric for Issue Presentation

3	2	1
Presentation clear and informative and peer teaching effective; preparation very evident	Presentation explained but lacked insight and enough preparation for teaching	Presentation not clearly explained; lack of preparation noted
Good selection of items used to support presentation; very interesting; well organized	Selected items were appropriate but presenters did not always seem interested in teaching	Selection of items presented not interesting and little effort shown in preparation
Issue clearly explained and listener gains insight	Issue explained well enough for listener to gain some insight	Issue not explained clearly enough to give listener insight
Sharing and cooperation of group members very obvious	Some sharing and cooperation noted	Sharing and cooperation not obvious
Displayed evidence of having used three or more different kinds of technology	Displayed evidence of using two kinds of technology	One or fewer kinds of technology used
Closing included information and supported conclusions	Closing stated but conclusion did not have much information to support it	Conclusions were not stated
Wrote and sent a thank-you to resource people		
Comments:	Total Points _____	

Table 4.9 Rubric for Model Building to Illustrate Understanding of Electricity Concepts

3	2	1	0
• Item of excellent design • Economical and appropriate use of materials • Complete circuit	• Item of good design • Some economical and appropriate use of materials • Complete circuit	• Item poorly designed • Wasteful or inappropriate use of materials • No complete circuit	• No item designed

(Continued)

Table 4.9 (Continued)

3	2	1	0
• Picture drawn and completely labeled	• Picture drawn but not labeled	• Picture drawn but not labeled	• No picture drawn
• Materials list complete and practical	• Materials list complete but not practical	• Materials list incomplete	• No materials list
• Model built and works well • Model neatly done and attractive	• Model built but doesn't fully work • Model lacks some attention to detail	• Model built but doesn't work • Model is sloppy	• No model built
Comments:			

Table 4.10 Rubric for Project to Design an Animal Theme Book

Group Report on Book With an Animal Theme		
Great Work	*Acceptable Work*	*Less Acceptable Work*
Tells about the book; includes most significant details without repeating word for word; listener would be able to retell with meaning	Able to relate some significant details; left out some important details that would help others better understand the story; listener may not be able to completely follow the story line	Tells some things about the story but details are unrelated or may be insignificant; listener would not be able to repeat the story
Able to rewrite information in own words appropriate to the theme of the book	Able to rewrite information in own words but left out some details important to the story line	Unable to rewrite; copies from original work; merely repeats what was already written
Able to draw illustrations of animals; reflects content of the book's theme	Able to draw illustrations of animals; may not all reflect the book's theme	Unable to complete illustrations or illustrations do not support book's theme

Group Proposal for New Book With an Animal Theme		
Great Work	Acceptable Work	Less Acceptable Work
All took active part in decision making; followed rules and showed courtesy toward others' ideas and actions	Some took active part in decision making; some disruption in the group	Members did not participate in decision making; did not use time to develop proposal
Logical suggestions made for animals to be included in new book	Suggestions needed more clarification but developed a proposal	Some attempt made to develop a proposal but didn't complete task; work needed in group dynamics
Comments:		

A Rubric for Machine Design

Task Description

Imagine that you work for a company that develops machines that can make everyday chores a little easier. You have just been asked to design and make a machine that combines two or more simple machines to complete one job (Table 4.11). You will be expected to complete a drawing of your machine and present at least a 1-minute commercial demonstrating your invention.

Table 4.11   Rubric for Machine Design

Excellent	Good	Improvements Needed
Two or more simple machines used in combination to perform a job	Only one simple machine was used, or two simple machines used but not clear how the two worked together	Simple machines were not used, or student did not appear to know how to complete the task
Invention appears to function in the desired way; realistic proposal	Invention works as shown but is not very realistic	Invention does not perform desired function
Simple machines described in a way to show energy conservation (savings)	Simple machines not described well enough to consider energy savings	No description of energy savings
Drawing completed; clear, neat, labeled, detailed	Drawing completed but lacks some details	Drawing completed but quality needs to improve; no attempt made
Commercial clever and creative; well planned	Commercial lacks some creativity; some planning but a few revisions needed	Commercial attempted but more planning needed; no attempt made

A Rubric for Construction Plan of an Ice-Fishing Shelter

Task Description

A third-grade class has studied the four seasons and how the seasons affect personal activities.* During the winter, some of the students would like to go ice fishing. Richard has agreed to build a shelter for them, but the students must develop a plan for what they want built. Mrs. Morrison insists that the shelter be as warm as possible because she gets really cold in the winter. Please do not include anything that may melt the ice because Mrs. Morrison cannot swim and does not want to learn how in cold water.

Draw and label a plan that Richard can follow to build this shelter. Include instructions about materials that are to be used and the dimensions that are to be used. Last, include a paragraph explaining why your plan is the one the class should present to Richard. Give reasons that will convince us that you have developed the best plan (Table 4.12).

Table 4.12   Rubric for Construction Plan of an Ice-Fishing Shelter

3	2	1
Good selection of materials. They will do the job and provide insulation from the cold; reasonable comfort and durability.	Information about materials is provided but may not be the best materials for the job.	Does not provide information about materials used.
Dimensions are complete and reasonable. A good shelter should result.	Dimensions are given. May be unreasonable for the job or may be incomplete.	Doesn't give dimensions for the shelter.
Paragraph contains sufficient and good reasons why your plan is best. Should be accepted by the builder.	Paragraph explaining why plan is best is provided. May not be backed up with facts.	Paragraph describing why your plan is best is missing or superficial.
Project is clearly written and easy to understand. Attractive presentation causes builder to choose your design.	Project can be understood but lacks attractive presentation.	Project is messy and difficult to understand.
No errors in computation to interfere with successful completion of the shelter.	Enough errors in computation to cause finished product to turn out incorrectly.	Major errors have been made in computations.

---

* This task might be modified since ice fishing is not common in some temperature zones! For example, students could design a shelter for a pet.

HOW WELL DOES A RUBRIC COMMUNICATE THE ORIGINAL TASK?

Table 4.13 was designed to assess a wildlife project that students were to complete. The rubric should communicate the major expectations for the tasks that students are to perform, and from the rubric, the reader should at least be able to construct the framework of the original task. The rubric should also send a message about what is valued in the performance. Can the original task be constructed from this rubric?

Table 4.13   Criteria for a Wildlife Project

Criteria	5	3	1
Appropriate season and climate	States in which seasons your projects will be used, including why they are appropriate	Seasons stated but appropriateness not stated	Invalid seasons and climate
Appropriate plants and animals for biome	Project appropriate for the animals and plants in your area	Project appropriate for either animals or plants in your area	Project is invalid for animals and plants in your area
Description of biome food web	Describes how your project benefits your area, including a six- to nine-organism food web	Description of benefits and four- to five-organism food web	Description of benefits and less than four-organism food web
Description of environmental maintenance	Description of how your project will maintain the balance of nature	Description of only your organism, not of surrounding environment	Description invalid for organism or environment
Presence of basic needs of organisms	Project adequately supports at least one basic need of organisms	Project inadequately supports the basic needs	Project does not support a basic need

Bottom Line: Not Every Rubric Is a Good or Useful Rubric

The rubric in Table 4.14 is a work in progress and is in the "not yet acceptable" category. Consider changing some of the wording. Does the word *describes* convey something different than the word *explains*? Are aesthetics important? The rubric conveys that aesthetics are valued, but what is beautiful to the teacher may not be so to the student. Other categories need work as well. What is *cramped*? The purpose of including this example is that it is a start, but it serves as a good example of a rubric that is not very useful for assessing a portfolio. If your classroom uses rubrics, revisit your goals and objectives and structure your rubric around these.

Table 4.14   Portfolio Assessment Rubric: An Example of a Not Very Useful Rubric

Criteria	4	3	2	1
Explains what my successes are	Clear, complete, some creativity to clarify	Clear, complete	Complete but unclear	Sketchy, vague
Explains what is presently being done in my classes	Gives specifics and clearly identifies major parts of each unit	Gives specifics of all units or projects	Gives some specifics	Only vague ideas
Pleasant to look at	Beautiful	Nice	Average	Plain
Easy to read	Clear, correct, fun to read	Clear and correct	Clear	Fairly clear
Organized	Well laid out with neatness	Neat, no direction	Cramped	Cramped and messy

A Second Example of a Not Very Useful Rubric

The rubric in Table 4.15 could be the template for a rubric to assess a laboratory performance related to the effects of gravity, friction, and mass on motion. From the rubric, it appears that this may be some type of culminating assessment because the effects of three variables are being considered. Some problematic words exist in this rubric. What is meant by *accurately?* What does *relate* or *related* mean? Nothing about the level of conceptual understanding is really conveyed. It sounds as if a textbook definition would also do the job, and that is probably not what the rubric developer had in mind. Rubrics take revision and refining. The Motions Concepts Rubric is in the "not quite there yet" stage.

Table 4.15   Motion Concepts Rubric

5	4	3	2	1
Accurately computed speed		Attempted formula but miscalculated		Made measurements and attempted computations
Accurately explained the effect that gravity has on motion		Partial description of gravity		Unable to relate gravity to motion

5	4	3	2	1
Accurately explained the effect friction has on motion, gave an example		Gave partial description of friction or gave an example of friction		Unable to relate friction to motion
Accurately explained the effect mass has on motion		Gave partial description of the effect mass has on motion		Unable to relate mass to motion
Accurately related three laws of motion		Accurately related two laws of motion		Accurately related one law of motion

A Third Example of a Not Very Useful Rubric

Suppose students were asked to follow the rubric in Table 4.16 in preparing an essay on ecology:

Table 4.16   Rubric for Essay on Ecology

Criteria	3	2	1
Cover	Attractive, colorful	Colorful	No color
Essay presents a lesson	Clear and creatively presented	Stated clearly	Not clear
Supports the lesson	Vivid details, understand	Little content	No content
Grammar	Correctly presented	Presented one to four errors	Many errors
Use of capital letters	Correctly presented	One to three errors	Many errors
Illustrations	Clear, colorful	Clear	Not complete

The message conveyed by this rubric seems to be that the essay is to focus on writing mechanics. This seems counter to the title, "Essay on Ecology," which appears to indicate that some criteria about understanding selected ecological concepts might have been expected. If the essay is going to be scored on writing mechanics, a separate score for these elements could be awarded and a second rubric could be used to assess conceptual understanding. Also, no context, other than ecology, is provided with this rubric.

5

Science Notebooks

Science notebooks have become a popular tool in many classrooms. This chapter addresses some ways that notebooks can be used to support students in learning science and sets out some ideas for structuring notebooks and assessment strategies. Notebook work also has the potential to infuse all six of the science domains.

SCIENCE NOTEBOOKS: FORMATIVE AND SUMMATIVE ASSESSMENT TOOLS

Scientists and researchers typically keep a notebook or log of their work for documentation and support for findings, and the nature of these notebooks may vary depending on the kind of research being conducted. As an example, the log of a field biologist would likely have data and notes different in format from those of a biochemist whose work focused on elucidation of molecular structures. Science notebook usage in the K–12 sector has been adopted by teachers in numerous classrooms. The use of science notebooks by students to write and record science work, completed or in progress, has become a source of both formative and summative assessment information. Success with science notebooks requires both planning and scaffolding to assist students in their efforts to keep a notebook as evidence of their learning. Based on his school district's success with science notebook implementation, Klentschy (2008) provided numerous suggestions about how to use science notebooks.

For the purposes of formative assessment, notebooks can serve as a context for giving students feedback and are records that provide evidence of student learning. Misconceptions can be identified in the work samples, and multiple opportunities are present for self- and peer assessments. Organizational skills are also reinforced. Science and literacy overlap, and research evidence is building that this kind of work in science helps to improve scores on some standardized measures. Student work with notebooks has a positive carryover in language and writing scores as well. Students have direct experiences and are able to write about things they can actually touch and see. Reflective writing helps students synthesize data and processes they have learned. The student is engaged in self-dialogue, and a dialogue with the teacher is also promoted. Students have opportunities to construct understanding while they draw, describe, create charts, and reflect upon their work. In *Linking Science and Literacy*, Douglas, Klentschy, and Worth (2006) provided examples that illustrate the benefits of notebook use. Science notebooks become tools for students to help explain their thinking and for justifying their ideas based on the evidence gathered in science work.

Science notebooks can also be a summative assessment of student mastery of content and process. They can serve as a tool for increasing student learning and for planning future instruction. They can become a bound document of reflections and communications of student understanding, and they can be effective substitutes for black-line masters. The notebook becomes a personalized document, and a level of student ownership develops that is not fostered with a worksheet approach in science. With student and parent conferences, the notebook becomes a powerful context for a discussion focused on student work samples.

If you are not convinced about the notebook approach, try beginning with a unit notebook. Even for younger students, they can begin by drawing and writing. A class notebook with work from each student or from student groups could be compiled and shared with parents. Not only does this document what a class is doing and learning, but this kind of work can promote a sense of student ownership. When a student work product is generated, student investment can be a powerful outcome that is not necessarily created by the completion of just another worksheet.

Format for Science Notebooks

The format selected can vary depending on the age of the students and other factors. In some classrooms, teachers opt for inexpensive composition books; students can begin with Volume 1, Issue 1, and if Volume 1(1) becomes filled, students can move on to Volume 1, Issue 2, in the same way that periodicals are sequenced. Composition books with quadrille paper work well when students are in grade levels where data collection and representation require graphing. Nothing is magic about the container for student work, but the dedicated notebook seems to generate a sense of permanency that a folder or three-ring binder does not. In research laboratories, the laboratory notebook may still be a carbon-backed version. As a researcher, you are expected to enter data and keep dated records. You do not erase, but instead draw a line through an entry that is in error. You do not tear out pages! Laboratory notebooks also become the bases of documentation for patents—so good records are important and invaluable.

Teachers often have a designated storage place in the classroom where students can leave their notebooks. Teachers who manage multiple sections can color-code notebooks by using a colored marker to draw a stripe across the edge of the notebook pages. The composition-size book is also convenient for younger students to manage when the notebook is used outside the classroom.

NOTEBOOK SCAFFOLDS

Notebook Organization

The organizational features of a notebook are a personal choice. Middle school students in a focus group session noted that one thing they found useful about notebooks was that they helped them stay organized. Students actually liked keeping notebooks. If notebooks are going to foster science learning, a recommendation that helps to do this is to provide some scaffolding for usage.

Page Format

Decide on a placement for page numbering and date. Do you want students to write on only one side of the page, or is front and back okay? If only the right side is used, students can use the left side to add notes and make annotations, and feedback can also be provided on the page.

Pen or Pencil

This choice may depend on the age of your students. In research labs, waterproof ink is a good choice. You do not want your records washed away! As noted earlier, if edits are needed, have students draw a line through the work that needs to be changed.

Table of Contents

Teachers often assist students by creating a table of contents on a sheet of chart paper. The teacher adds to the table of contents each day to guide the students in doing so, and this becomes a running record of daily science activities.

Notebook Entries

What should be written, drawn, and included in the notebook? This, of course, depends on the developmental level of students. A recommendation for any level is to begin with a focus question for the day. Communicate the expectations for learning to students. Writing should become an act that is automatically embedded in what students are doing. Save paper; instead of a handout, have students enter the information directly in their notebooks from a template that is projected. A group copy can be provided, and students can transfer the information to their notebooks. If you have handouts for students, orient the information in landscape format so pages can then be cut in half. This makes adding the page to the notebook easier. Students can tape, paste, or staple the half-page into their notebooks, and with this approach, the work on notebook pages is visible when looking through the notebook. Reviewing a notebook where every sheet needs to be unfolded to be read does not appear to be thinking made visible! With increasing copy costs, the half-sheet approach is more economical if students need copies.

At the end of a class, have the students summarize what they have learned in a sentence or two. As the teacher, return to your focus question and ask students to write a sentence or two about the science they have learned. Some teachers call these *lines of learning*; in addition, as closure, students can also be asked to write down a question they have after the lesson. This question might be about something they did not understand, or it could be an "I wonder" type question. Students could also write these kinds of questions or lines of learning on an exit slip. Students can also be asked to use specific colors to highlight the "what they have learned" statements and questions that they have. Information can be color-coded with highlighters to help students and teachers more readily find and return to ideas. Using sticky notes of various colors for tabbing information that the teacher wants to check is another way teachers can manage checking and responding to work in notebooks.

When students are working and recording information in their notebooks, they can be asked to write a question mark in the margin next to something that they do not understand. Encouraging students to do so can help them to self-monitor and self-regulate their personal learning. This also supports the kind of classroom climate where we as teachers expect students to have questions.

The lines of learning and questions that students have also make good entry points for the next lesson in a sequence. These can serve as a quick assessment to see what students are learning and understanding. Also, returning to the questions sends students the message that you use and value what they are telling you. Student ownership comes into place when they are communicating what they have learned and are not being told what they have learned by the teacher.

Chesbro (2006) suggested using what he called an *interactive science notebook approach*, which has commonalities with reading-response and double-entry approaches that are used in English, writing, and reading. With this format, the right page is viewed as an input page on which incoming information, such as notes, is recorded, and the left page is an output page where the student sets out information that connects to the information on the right page. Students also personalize this information in ways that illustrate what they are learning.

Model Entries With Students

When you are beginning work with notebooks or with a new group of students, it will help students to know what is expected of them in formatting and recordkeeping. This should help students produce better-quality work, and this can make the task of responding to student work more manageable. If available, work samples from prior classes can be used as exemplars. Student names should be removed, and pages can be scanned for use with other classes. The act of keeping a notebook should become a habit for students, and without asking, they should internalize the process as a routine expectation in the classroom.

ASSESSMENT COMPONENTS FOR NOTEBOOKS

The *National Science Education Standards* (National Research Council [NRC], 1996) and the *Classroom Assessment and the National Science Education Standards* (NRC, 2001a) call for assessment emphases that focus upon the following:

- Assessing what is most highly valued
- Assessing rich, well-structured knowledge
- Assessing scientific understanding and reasoning
- Assessing to learn what students understand
- Students engaged in ongoing assessment of their work and the work of others
- Teachers involved in the development of external assessments

Except for the last bulleted item in the list, science notebooks align with these recommended areas. In fostering student work in notebooks, teachers have a record of what students are learning and understanding. If a teacher does not see the kinds and levels of learning expected, this could be attributed to several reasons. The students may need more practice and instruction, or perhaps the lessons and approach to concepts might need modifications. From informally assessing the notebook work, the teacher should be able to determine whether most students do understand and if there are isolated problems. Gilbert and Kotelman (2005) set out five reasons to use notebooks, and they indicated that notebooks are first of all thinking tools to support conceptual learning. In reviewing notebook work, teachers can use the student work to guide instruction, and in addition, notebooks can enhance the development of literacy skills.

With notebook implementation, a record of work is created over time, and this record is a context for ongoing formative assessment. With formative assessment, three things are required to promote learning: a clear view of the learning goals, information about the present state of the learner, and action to close the gap (NRC, 2001b). Use of the lines of learning serves as a daily checkpoint that can assist in doing these things. Shepardson and Britsch (1997, 2000) also noted that when examining students' science journals (notebooks), it is necessary to carefully inspect the drawings and writings and really assess the levels at which students understand. Shepardson and Britsch pointed out the importance of determining if the students' work represents learning science and not just learning an activity. Does the student work provide evidence of conceptual understanding?

When students are making observations, having them write and draw what they notice supports development of multiple skills. If students have limited writing abilities, they can verbally describe what they draw, and vocabulary development can be supported for all students. Looking for and noting details supports conceptual development in science. Development of expertise comes with focus and study, and scientists even talk about becoming one with an organism when they have studied and researched it over time. Observational skills can be developed, and observational powers provide sources for building analogical thinking. Drawings also provide a context for going back and revisiting concepts. Edwards (1979) stated that drawing is a curious process that is so intertwined with seeing that the two can hardly be separated.

Differentiated instruction, which can help to narrow achievement gaps, is also supported with a notebook approach, and notebooks foster collaboration not only within the classroom among students but also among teachers whose classrooms use notebooks. The notebooks are a source of student work that can be examined by teachers for professional development. From evidence in student work, teachers can determine what is working and can also find areas that need improvement. With any of the process skills, working with a notebook supports development in all areas. Students can support their thinking with notes, drawings, written plans, and records. The notebooks can become the "story" of science for the year.

Notebooks can also be used for summative purposes. Across a series of units, the notebooks can be assessed for conceptual development, for increasing development of student skills and abilities with process skills, and also for working in ways that reflect the scientific endeavor. Students can also be involved in self-evaluation of their work and personal knowledge growth.

Teachers who use notebooks find that scaffolding an investigation with a checklist framework (Figure 5.1) helps students remember to address each component in an investigation. Figures 5.2 through 5.5 provide ideas for ways to use notebooks for self- and peer assessment and also for communicating with

Figure 5.1   Science Notebook Components for an Investigation

Place a colored tab by each component in your investigation.

Question/Problem/Purpose	(color code)
Hypothesis Prediction (younger students)	(color code)
Plan for Investigating Question	(color code)
Observations/Data/Results	(color code)
Evidence-Based Conclusion Based on my data I learned . . .	(color code)
Next Steps/New Questions	(color code)
(Other)	(color code)

Figure 5.2   Checklist for Science Notebook Components

Notebook Components	Student Self-Check I've done this.	Teacher Check You've done this.	Comments
Question/Problem/Purpose			
Stated in my own words			
Related to purpose/big idea			
Clear and concise			
Investigable			
Prediction			
Connected to prior experience			
Clear and reasonable			
Related to question			
Gave an explanation/reasons			
Planning/Procedures			
Related to the investigable question			
Identified variables/control			
Included data organizer—knew what kinds of data to record			
Stated materials needed			
Clear sequence/direction			
Modified as needed			
Gave enough detail to repeat			
Observations/Data/Results			
Related to question and plan			
Included drawings, charts, and/or graphs; wrote about what was done			
Organized—could follow work			
Accurate—kept good records			
Conclusion/What I Have Learned			
Stated in my own words			
Clear statement of what I learned			
Based on question/planning/evidence			

(Continued)

Figure 5.2 (Continued)

Notebook Components	Student Self-Check I've done this.	Teacher Check You've done this.	Comments
Supported conclusion with evidence collected			
Reflective			
Showed rigor in thinking/recognized limitations of investigation			
Next Steps/New Questions			
A new question that I have is asked			
Extension/new application of the original question			
Researchable or investigable			
My thoughts/things I wondered about			
Raised and recorded questions throughout investigation			

Student Note to Teacher:

One thing that I learned about science from investigating this question was:

This is something I would like you to know about this work:

Figure 5.3   Student Self-Assessment of Notebook Entry

Name: _____   Class Period: _____

Date: _____

Notebook entry (the entry for this self-assessment is):

_____

Place a check on the line if you have these for the selected notebook entry. Write NA in the blank if this does not apply.

_____ Question/Problem/Purpose

_____ Prediction

_____ Plan

_____ Observation(s)/Data/Results

_____ Conclusion supported with evidence

_____ What I have learned

_____ Next Steps/New Questions

_____ Other: <u>science vocabulary</u>

_____ Other:

_____ Other:

Circle the description in each category that you feel best describes the notebook entry being assessed.

	Excellent	*Good*	*Need to Do More*	*Help Needed*
Effort	I gave this my best effort. I really tried.	I gave this some effort. I could have tried a little harder.	I really did not put much effort into this. I need to try much harder.	I need help here. (I decided not to even try to do this.)
Organization	My work is well organized and easy to follow.	My work is organized, but I could do a little better.	My work needs some major organization.	I need help here.
Neatness	My work is very neat.	My work is good, but it could be a little neater.	My work is somewhat messy. I really need to work on neatness.	I need help here.

Please answer the following questions.

1. What does this entry tell about the science you have learned?

2. What do you think is the best part of your notebook entry?

3. What could you have done to make your notebook entry better?

4. What is something that you want to tell me about your notebook entry?

Developed by Sandra K. Enger and Rebecca McCoy.

Figure 5.4 Peer Review of Notebook Entry

Reviewer's name: _____

Class: _____

I am reviewing the notebook entry for (name): _____

Date: _____

Place a check on the line if you find these for the notebook entry. Write NA in the blank if this does not apply.

_____ Question/Problem/Purpose	_____ What I have learned
_____ Prediction	_____ Next Steps/New Questions
_____ Plan	_____ Other: <u>science vocabulary</u>
_____ Observation(s)/Data/Results	_____ Other:
_____ Conclusion supported with evidence	_____ Other:

Circle the description in each category that you feel best describes the notebook entry being reviewed.

	Excellent	Good	Need to Do More	Help Needed
Effort	The student gave his/her best effort and really tried.	The student gave this some effort but could have tried a little harder.	The work looks like little effort went into this. He/she needs to try much harder.	Help is needed here (or, Did not even try to do the work).
Organization	The work is well organized and easy to follow.	The work is organized, but this area could be better.	The work needs some major organization.	Help is needed here.
Neatness	The work is very neat.	The work is good, but it could be a little neater.	The work is somewhat messy. He/she should work on neatness.	Help is needed here.

Please answer the following questions.

1. Did the "What I have learned" section relate to the question at the beginning of the lesson?

2. Did the "What I have learned" answer to the question make sense to you?

3. What was the best part of the "What I have learned" section?

4. How could the student improve his/her notebook entry?

Developed by Sandra K. Enger and Rebecca McCoy.

Figure 5.5   Parent Review of Science Notebook

Student: _____

Parent: _____

Please take a few minutes to look over your child's science notebook. I would appreciate your feedback on the work in the notebook. Please use the following guide, and circle the description in each category that you feel best describes your child's work.

	Excellent Work	*Good Work*	*Needs Some Work*	*Under Construction*
Science Content	I can really see the science. I can really tell what my child is studying and learning.	I can see the science, but I do have a few questions about what my child is studying and learning.	This looks like science, but from the work, I am not certain what my child is studying and learning.	Needs to work on this.
Student Effort	Excellent effort in all work.	Good effort in most of the work.	Some effort, but could do more.	Needs to work on this.
Organization	Very organized throughout.	Good organization. A few areas need a little work.	Some organization, but needs to include all information.	Needs to work on this.
Neatness	Very neat throughout.	Neat, but a few areas need some work.	Could work on this throughout.	Needs to work on this.

Please answer the following questions.

1. What do you like about the science notebooks?

2. Do you think it is helpful for your child to keep a science notebook? Explain.

3. What do you think your child can do to improve his/her science notebook?

4. Please write a positive sentence or two to your child about the notebook work that you reviewed.

Developed by Sandra K. Enger and Rebecca McCoy.

parents. When a different color code is set up for each component, notebook management is made easier. A sticky note can be placed by that notebook component, and that component can be easily identified for checking. When students are first learning to set up an investigation, the teacher may want to check that the question posed is one that actually can be investigated. Responding to student work and providing feedback are vitally important, and these are very time-consuming. Providing feedback in manageable increments helps keep the task of responding updated and supports the notebook process. With attention to individual components and some modeling, students are better able to produce quality work and design investigations.

In Figure 5.6, the point value assigned to each category is arbitrary and can be adjusted to align with a teacher's assessment plan. If you have students write lines of learning and questions they have after lessons, you may want to revisit some of these when you assess notebooks. In a preservice science-methods class, I ask students to add an entry at the end of the class session. This entry is called a "Science Teaching Line of Learning" (STLL): *After today's class, I understood that....* I also ask them to add a second entry called "Concept Questions" (CQs): *A conceptual question that I have is....* Students are asked to highlight STLLs in yellow and CQs in orange. The STLLs and CQs can be discussed at the end of class or used as lesson starters for the next class. Students are also asked to word process and submit these. If you want to know if students are developing science understanding, the lines of learning can provide an assessment checkpoint at the end of a class lesson. This assessment is embedded in your instruction, and this assessment should inform your instruction and guide the next steps in instruction.

SUPPORTING GRAPHING SKILLS AND ABILITIES

Data representation and interpretation are areas that present challenges for many students, and these are also skills and abilities that cross disciplinary boundaries. Items requiring reading charts, tables, and graphs are typically included on standardized assessments.

Generally, the purpose of producing a graph is to communicate information in a pictorial mode. A graph should tell the story of what was done in an investigation or study. This goes to that phrase that we all know—*a picture is worth a thousand words*—or in this case, perhaps a thousand numbers. Examples of graphs that can be used for practice with interpretation can be found online, in newspapers, and in periodicals.

With younger students, the use of manipulatives supports the development of graphic depiction. Developing fluency with graphic representation requires spatial thinking as well as some level of abstraction.

To help develop fluency with graphing, students need to have practice with the development and preparation of graphs. The skills of the process should be explicitly taught, and student discussions centered on graphs that exemplify attributes of good practice help students to fine-tune their abilities. Once students are familiar with the expected attributes, having them critique graphs that lack these attributes is useful in their work toward skill

Figure 5.6   Self-Assessment for Science Notebook Work

Self-Rating (Please Circle)	*Notebook Component*
(20) A B C	Level of notebook keeping: • I kept detailed notes and records. I made extra efforts to add extra details, and my records are complete and easy to read and follow. Holistically, I would say that I earned an "A." • I kept adequate notes and records. I really could have put forth greater effort in completing notebook entries every class meeting. My records are quite neat. I would say that I have earned a "B." • I tended to let my notes and records go a bit. Sometimes I did not expend a great deal of effort in writing things down. The level of neatness is adequate. I would say that I have earned a "C."
(20) Top-quality work Some improvements needed Major improvements needed	Observations and Drawings/Diagrams • What I did well and why this was done well: • What I need to improve upon: Use a Post-it to tab at least three (3) examples of your top-quality work. (On the Post-it, write a brief note about why you consider this top quality.)
(5) A B C	Table of contents: • I kept an updated table of contents and was consistent about doing so. Anyone could easily locate daily work, observations, drawings, etc. in my notebook. I would say that I earned an "A." • I slipped some days so I would say that I earned a "B." • I had problems remembering to write things down and need to work on consistency. I would say that I earned a "C."
(5) A B C	Pages numbered and dated: • I kept up with dates and pages, and I was consistent about doing so. Anyone could easily locate daily work, observations, drawings, etc. in my notebook. I would say that I earned an "A." • I slipped some days so I would say that I earned a "B." • I had problems remembering to write things down and need to work on consistency. I would say that I earned a "C."
(10) A B C	Based on the overall quality and level of my work from a holistic perspective, I would rate my science notebook work as "A" quality, "B" quality, or "C" quality based on the following justification:
Instructor's Response Comments:	

Developed by Sandra K. Enger.

development. Whatever the work, it should be developmentally appropriate for students and connected to science contexts that are the foci of the concepts that students are learning. Graphing calculators and software such as Excel are excellent tools, but some paper-and-pencil practice can provide knowledge and insights into how and why graphic information is presented and communicated in certain ways. InspireData, a user-friendly software package, is available online (http://www.inspiration.com). Templates to generate graph paper are also available online (from sites like http://sme.clc.uc.edu/graphpaper.htm).

In science, when we work with students to practice and learn graphing skills, we often do so in the context of some experiment. One such experiment could be an investigation of the relationship between the length of the pendulum and the number of times a pendulum of constant mass swings in a given time interval. Setting up an experimental design for an investigation is a context in which direct instruction and scaffolding are productive methodologies. A checklist can provide a scaffold to remind students about what should be included (Figures 5.7 and 5.8).

Figure 5.7 Checklist for Graphing

Check		Criteria
	1.	Type of graph selected is appropriate for data. I chose a _____ (type of graph) because:
	2.	The graph has a title.
	3.	I have provided a key or legend.
	4.	I have labeled all parts of the graph.
	5.	I have used an appropriate scale.
	6.	I have made an accurate graph.
	7.	I have a graph that is neat.
	8.	I used the necessary tools to construct the graph (ruler, compass, etc.).
	9.	If I gave the graph to someone, it clearly shows what I have done.
	10.	Other:

Figure 5.8 Checklist for a Line Graph Plotted on *X-Y* Coordinates

Check		Criteria
	1.	I am using a line graph because:
	2.	I gave the graph a title.
	3.	I have provided a key or legend.
	4.	I have labeled all parts of the graph.
	5.	I have used an appropriate scale.
	6.	I have made an accurate graph.
	7.	I have a graph that is neat.
	8.	I used the necessary tools to construct the graph (ruler, etc.).
	9.	If I gave the graph to someone, it clearly shows what I have done.
	10.	I have plotted the manipulated variable (independent variable) on the *x*-axis, and the responding variable (dependent variable) on the *y*-axis.

Students develop at various rates in relation to their cognitive abilities and formal operational thinking, and work with experimental design is typically a part of curricula in fifth grade since that is the age when students are more likely to be developmentally ready for the kinds of thinking involved.

6

Assessment Examples for All Grade Levels

In this chapter, you can find suggestions for the following:

- Opportunities for student self-assessment in group work
- Assessing students' resource use for projects or assignments
- Using open-ended questions
- Using the Student Laboratory Environment Inventory (SLEI)
- Assessing in the application domain
- Using an observation checklist
- Using activities to foster creativity
- Using application-level multiple-choice questions
- Encouraging analogical and metaphorical thinking
- Assessing attitudes, preferences, and processes, including the STS Attitude Scale

Engaging students in assessment of both their own work and their performance in a group can assist them in seeing their roles in the accomplishment of group tasks. Awareness of personal contributions can help students internalize criteria and also likely assists in the development of metacognitive abilities. Several suggestions for student self-assessment in small-group work are presented in Figures 6.1, 6.2, and 6.3.

An example of the reliance on students' self-evaluation is shown in the resource use inventory (an example is shown in Figure 6.3). This inventory can help provide focus for work in which student use of information beyond the textbook is desired. Making connections via accessing a wider range of materials can encourage greater student independence.

Figure 6.1 Student Self-Assessment: Group Evaluation (Attitude Domain)

Group activity: _____

Student name: _____

For Questions 1 and 2, circle the words that describe your thinking about how your group worked together on the activity.

1. Overall, how well did your group work together?

   Very well   *Quite well*   *Not very well*

2. Overall, what rating would you give your group's product (poster, pamphlet, skit, etc.)?

   Very good   *Quite good*   *Not very good*

3. What do you think was the best part of your group's product?

4. What is one thing that your group could have done to improve your product?

5. What is one suggestion you have for helping groups work together?

Figure 6.2 Individual Student Self-Assessment (Attitude Domain)

Name: _____

For Questions 1 through 3, circle the words that describe how you think you worked with your group.

1. How much did you contribute to the group product?

   More than others   *Same as others*   *Less than others*

2. Did you offer ideas?

   More than others   *Same as others*   *Less than others*

3. Did you accept ideas from others in the group?

   More than others   *Same as others*   *Less than others*

4. What is one thing that you would like others to know about your work on this product?

Figure 6.3   Finding Information for Class-Related Work (Application Domain)

Read each of the following questions, and then check (√) the column that best describes your use of the information resources. If you have never used or do not do this, put a check in the last column.

Finding Information for Class Work	Almost Every Day	At Least Once a Week	Hardly Ever	Never Use or Do This
1. Do you use the Internet to help with class work?				
2. Do you use Google, Yahoo, or other search engines?				
3. Do you use email to contact and/or communicate with others about information?				
4. Do you blog about your work?				
5. Do you listen to or watch podcasts or media clips for class work?				
6. Do you use materials from the school library or media center for class work?				
7. Do you use newspapers, magazines, or journals to help with class work?				
8. Do you use a computer for class work?				
9. Do you use information from TV or radio to help with class work?				
10. Have you used materials from sources other than those already mentioned for class work?				
11. Do you ever discuss science topics with adults?				
12. Do you take notes to help you remember ideas?				
13. Do you take notes when doing work for class-related work?				
14. Do you take notes without being told to do so?				
15. Do you ever contact resource people for information related to class work?				

16. What is your favorite way to find information for class work?

17. How do you determine that the information you access is credible?

OPEN-ENDED QUESTIONS (INTEGRATES DOMAINS)

Students can respond to open-ended questions by providing explanations, opinions, and solutions to problems. The responses can reveal students' ideas and conceptions much like an individual oral interview (Strauss & Stavey, 1983). A variant of open-ended questions requires students to provide definitions. The interpretation of responses does then require some caution because students can memorize definitions by rote (Ausubel, 1968). However, incorrect definitions can sometimes reveal misconceptions (Barenholz & Tamir, 1992).

An open-ended question will not have only one possible response, and students should be expected to have a reference listing to support proposed answers. Issue-oriented contexts can serve as sources of ideas since most issues tend not to have one right answer, but solutions offered must have support. One solution may be the best candidate at the time. In the following examples, a context is given, a question follows, and students are asked to explain.

Example:	People are able to live healthy lives eating only fruits and vegetables and no meat. Explain how this can be possible.
Possible Explanation:	No one fruit, vegetable, or meat contains all the necessary nutrients for healthy living. A vegetarian diet can be eaten that provides the nutrients essential for good health. A vegetarian would have to be certain to include nonmeat foods that provide the proteins supplied by meat.

Possible Contexts for Students to Address

Students could first be given the question of how the caloric intake of a vegetarian compares with that of a nonvegetarian. To further initiate their thinking about caloric intake, students could be asked to respond to the following questions:

- How do the numbers of calories in 10 grams of bread, lean hamburger, and celery compare?
- Use a graph to show the number of calories available in 10 grams each of bread, lean hamburger, and celery.

  Performance Expectation: The students should draw a graph, provide an appropriate title, and record the units. The results are likely to show that the caloric values of meat and bread are very similar, whereas celery has a much lower caloric value.

- How have school lunch menus been modified in response to recent health concerns related to obesity? (What would you need to know to determine if current school lunches were nutritious and balanced?)

  Performance Expectation: School lunch menus, portion sizes, and food preparation details need to be analyzed. The chemical composition of the lunches; the fat, carbohydrate, and protein content; their caloric values; and the relative amounts of nutrients need to be considered in the answer. Current lunch menus could be compared to past menus.

- The President's Challenge for Physical Activity and Fitness provides tools and background information for various age groups. What are the recommendations for your age group? Once you have identified these recommendations, design a fitness plan for your class.

  Performance Expectation: Students need to investigate the recommendations for the President's Challenge that are available online. Once they have studied the recommendations, they can then design a fitness plan that could be submitted for review and potentially implemented.

- Environmental issues are very applicable to open-ended questions or problems. The Environmental Protection Agency (EPA) (http://www .epa.gov/epahome/home.htm) has multiple topics for consideration. Endangered and threatened species are often of interest to students. What can be done to save species in your state or area? When animals, such as wolves, are reintroduced, what problems are solved? What problems are created? What problems do invasive species create for native populations? Many other possibilities exist for students to investigate questions and propose plans that offer potential ameliorations.

  Performance Expectation: Depending on questions formulated, students could investigate the underlying factors that generated a problem or concern. They could design a proposal for review and possible implementation. A town hall meeting or debate format could be used for discussion of the question investigated.

- Health issues represent another context replete with open-ended questions or problems. The World Health Organization (http://www.who.int/en) could be a starting place for selecting problems. The Centers for Disease Control (http://www.cdc.gov) could also be a resource. Students could investigate questions related to the overprescription of antibiotics, whether to vaccinate or not, the outbreaks of disease, or many others.

  Performance Expectation: Depending on questions formulated, students could investigate the underlying factors that generated a problem. They could discuss the pros and cons of certain areas. Students could develop a position paper supporting their stance.

Teachers can also engage younger students with open-ended questions. What kinds of snacks are the healthiest to eat? Teachers could locate resources appropriate for the students, and students could develop a list of snacks that would be healthy choices. Ideas and suggestions at MyPyramid (http://mypyramid.gov) can provide a starting point. Where possible, students could plan the choices for snack time or identify items that would make healthy afterschool snacks.

Centered on weather events, younger students could discuss and develop plans for the kinds of actions they should take at home, in school, or when out with friends or family if an emergency should occur. Students could decide what they should have in an emergency kit and which places are best for taking shelter.

Knowing how to respond to emergencies is important for all ages, and with younger students, they could help develop a plan for action. What is 911, and when should you call 911? While this kind of activity is not necessarily open-ended, it helps children evaluate situations and make decisions.

Students at very young ages can also be engaged in projects that benefit the environment. Not littering and recycling are two societal projects that students can have ownership of right in the classroom.

Planting and growing a flowerbed or garden are also choices that work well for students of all ages. If there is an area on the school grounds where this can be done, it can promote school ownership. Raised beds or container gardens can fit into very small spaces. Students can plan their gardens, and mathematics can be incorporated. If students are going to garden, they will need to decide what and how to plant for the climate and available space. A budget could be developed, and tasks could be assigned. A butterfly garden could also be researched and developed.

When the forty-fourth president took office, a question arose about what kind of dog would be a good pet for his daughters. Students very often want pets or have pets, and questions could center on pet selection. What kinds of pets would be good choices for a classroom? What kinds of care do pets require? How much does it cost to keep a pet? Any number of questions could be pursued and investigated.

Although this section is titled "Open-Ended Questions," the kinds of products that students can potentially generate would align well with a project-based approach (Krajcik & Czerniak, 2007). A project-based approach features a driving question that provides the direction for the work that students and teacher undertake. A project-based approach interconnects and integrates the six domains: concepts, processes, application, attitudes, creativity, and nature of science. With technology, opportunities to collaborate on projects can even become international in scope.

STUDENT LABORATORY ENVIRONMENT INVENTORY (INTEGRATES DOMAINS)

Although much is known about hands-on learning, and many studies have corroborated that learning occurs best when actually doing science, the research has not been comprehensive in assessing the effects of laboratory instruction on student learning and attitudes. Laboratory experience is expected in the sciences, and the Student Laboratory Environment Inventory (SLEI) (Fraser, Giddings, & McRobbie, 1992) was designed to assess student attitudes about learning in the science laboratory. The SLEI is intended for use where a separate laboratory class exists, and this instrument may be more appropriate for upper secondary and higher education levels. The 35 items on the SLEI, with response alternatives of *almost never, seldom, sometimes, often,* and *very often,* measure five dimensions. Two forms of the instrument have been designed. An "Actual Form" asks students to note what actually takes place in the laboratory classroom, and the "Preferred Form" asks students to respond to what they would prefer to take place in the laboratory class.

Each of the aforementioned assessment activities can be completed in the classroom within the realm of authentic assessment that is embedded in the instruction. The students may be performing any one of these process skills during an instructional period, and the teacher could, using a predetermined observation checklist, observe the groups performing a particular class activity to determine a student's ability to perform that process. Figure 6.4 includes examples of items that could be used to assess some of the processes of science.

Figure 6.4   Checking for Process Skill Proficiency

Student: _____   **Date:** _____

Activity: _____

Process skill(s): _____

   3 = Student demonstrates proficiency

   2 = Student still needs more practice to become proficient

   1 = Student was unable to accomplish the activity without help

   NA = Not applicable for this activity or age level

Circle either the number or NA to indicate the student's level.

1. This student was using appropriate process skills to determine the outcome(s) for the activity.	3	2	1	
2. This student performed the process skill(s) within the context of the activity.	3	2	1	
3. The process skills used in this activity:				
observing	3	2	1	NA
classifying	3	2	1	NA
communicating	3	2	1	NA
measuring	3	2	1	NA
predicting	3	2	1	NA
inferring	3	2	1	NA
operationalizing definitions	3	2	1	NA
identification and control of variables	3	2	1	NA
generating and testing hypotheses	3	2	1	NA
designing experiments	3	2	1	NA
data interpretation	3	2	1	NA
developing models	3	2	1	NA

Notes and recommendations:

IDEAS FOR ASSESSING IN THE APPLICATION DOMAIN

Assessing students' learning in relation to their abilities to make applications involves asking them to apply principles or skills (or both) to new situations for which they are expected to provide unique solutions. Prior to assessing a student's ability to apply concepts, teachers should develop questions that promote the thinking involved in applications. Certain words and phrases can frame the thinking required in making applications.

Possible Prompts	Application Words	Application Words
What would you use to . . .	Separate	Predict
What would result if . . .	Arrange	Identify, classify
Tell what would happen if . . .	Select	Explain
Tell how much change there might be if . . .	Differentiate	Construct
	Demonstrate	Plan
Illustrate how . . .	Show how	Conclude
Choose the statements that apply . . .	Cause for	Diagram
	Compare and contrast	Outline
What is the reason for . . .	Investigate	Chart
	Dissect	Graph

Some examples of application questions follow, and these questions would typically be followed with further questions related to the how and why. Addressing the how and why can move students into using higher-order thinking skills.

Which of the following methods is best for _____?

What steps should be followed in applying _____?

Which situation would require the use of _____?

Which principle would be best for solving _____?

What procedure is best for improving _____?

What procedure is best for constructing _____?

Which of the following is the best plan for _____?

What is the most probable effect of _____?

Application questions that call for information use or skills often require original thinking because they may

- require knowledge of standard procedures;
- present a specific situation and require appropriate action, reasonable inference, or prediction;

- require solutions of numerical or other mathematical problems;
- present a statement or observation to be interpreted or evaluated; or
- assume unusual or complex situations and require inferences.

When constructing assessment instruments or items, the words and phrases used indicate the expected level of cognitive processing. The students should have the chance to integrate their procedural understanding as well as their conceptual learning to be able to respond to the questions. If the questions are designed to only assess student recall abilities, they will require less cognitive processing. Questions that ask students to combine processes and concepts they have been learning and then make applications are desirable. Students should have opportunities to use higher-order thinking skills, and questioning strategies can promote this construct.

OBSERVATION CHECKLIST (INTEGRATES DOMAINS)

The following example of assessment could be embedded within the instructional phase of a unit or module addressing such issues as flooding, ecosystems, wetlands, water pollution, weather, global warming, deforestation, urbanization, endangered species, or recreation.

Example: The students are given an activity in which two pertinent questions are asked. The students should form cooperative groups and be asked to investigate the questions provided in the activity. They may use any method on which they agree, and the solution must be in a form that can be presented to the whole class. The time frame for this activity can be decided on by either the teacher or by a student vote.

While the students are working on this investigation, the teacher may want to perform the embedded observational assessment provided in the following table:

Observed Behaviors	What Was Observed	Notes, Comments
Student groups are on task.		
Students are actively discussing issues.		
Students are using prior concepts to propose a solution or resolution.		
Students are exhibiting positive attitudes.		
Students are demonstrating an understanding of scientific problem solving.		

(Continued)

(Continued)

Observed Behaviors	What Was Observed	Notes, Comments
Students are using multiple strategies to address the problem.		
Students are using a variety of skills to present a group solution.		
Other		

This kind of checklist could be electronically supported or placed on an index card and carried by the teacher while circulating among the cooperative groups. Assessable components may include (a) students talking about their own experiences, thereby bringing prior knowledge into play (concepts may be overheard and shared by students); (b) communication skills within groups; (c) attitudes of individual group members or entire groups; (d) newly generated questions; or (e) creativity dealing with multiple variables. There are many more observations that can be noted depending on the prior criteria set up by the teacher and students.

USE OF APPLICATION-LEVEL MULTIPLE-CHOICE QUESTIONS

Multiple-choice questions can be written to assess concepts at the application level, and a search of test publishers' materials and state-assessment documents should also provide a source for these kinds of questions. Some examples are located in Figure 6.5; for each question, the best answer is indicated with an asterisk.

Figure 6.5 Sample Application Assessment Items (Application Domain)

The best answers are indicated with an asterisk.

1. *Concept:* When water freezes, its volume increases.

   Application question: Which one of the following is the main reason that water should not be stored in the freezer in a container that is completely filled and sealed?

   A. The taste of the water will change.
   B. The container might break as the water expands.*
   C. The water reacts with the glass at low temperatures.
   D. Water will not freeze if there is not enough space for it to change to ice.

2. *Concept:* The time to warm an object that is in a boiling liquid depends on the amount of material making up the object and how much of its surface is exposed to the boiling liquid.

   Application question: Which of the following examples will cook most slowly when placed in boiling water?

   A. A single 1-pound potato*
   B. One pound of small potatoes
   C. One pound of medium-sized potatoes
   D. One pound of potatoes cut into small pieces

3. *Concept:* Most bird identification guides are based on knowledge of a bird's shape, size, color, and patterns of markings.

   Application question: While sitting at the breakfast table on a winter morning, you notice a species of bird you have never seen before. Which one of the following would you recommend to best guarantee that you will be able to identify the bird?

   A. Make a note of the bird's favorite food.
   B. Observe the behavior of the bird.
   C. Study the size and coloration of the bird.*
   D. Determine the sex of the bird.

4. *Concept:* The temperature at which water boils decreases with altitude; therefore, the temperature of boiling water at high altitude will not be as high as the temperature of boiling water at sea level. A pressure cooker is a kitchen appliance where high pressure and high temperature are maintained inside the cooker despite the altitude.

   Application question: Where would it be most efficient to have a pressure cooker for cooking food?

   A. Below sea level
   B. At sea level
   C. At low elevations
   D. In the high mountains*

5. *Concept:* A great amount of heat energy is required to evaporate water (heat of vaporization); therefore, evaporation is used as a cooling process.

   Application question: If you were out on a camping trip, which of the following situations would result in providing the coldest drinking water? Assume that the amount and beginning water temperature are the same for each case.

   A. A metal canteen is filled with water and kept in the shade.
   B. A metal canteen is filled with water, covered with a wet cloth, and then kept in the shade.*
   C. A metal canteen is filled with water, immersed in a bucket of water at the same temperature as inside the canteen, and then kept in the shade.
   D. A metal canteen is filled with water and kept in direct sunlight.

6. *Concept:* Light-colored objects reflect sunlight better than dark-colored objects.

   Application question: On a sunny winter day, which vehicle would likely be the warmest to the touch?

   A. A blue car
   B. A red car
   C. A white car
   D. A black car*

7. *Concept:* In very cold conditions, up to 80% of the heat made by the body can be lost through the surface of the head and neck.

   Application question: If you were suddenly caught outside in below-zero weather and all you had to wear were boots, shorts, a T-shirt, and a light jacket, which of the following would be the best way to retain the greatest amount of body heat?

   A. Wrap the T-shirt around your bare legs.
   B. Leave the T-shirt on as you usually wore it.
   C. Wrap the T-shirt around your head and neck.*
   D. Use the T-shirt to keep the nearby air moving.

8. *Concept:* Objects with a large ratio of surface area to volume will cool faster than objects with a small ratio of surface area to volume.

   Application question: Suppose a waiter brings you a cooked steak. Which is the best way to keep the steak as warm as possible while you eat it?

   A. Cut only the piece to be eaten.*
   B. Cut the steak quickly into bite-sized pieces.
   C. Keep air moving near the steak.
   D. Eat slowly.

Figure 6.5 (Continued)

9. *Concept:* Warm-blooded animals lose body heat in proportion to their surface area and generate body heat in proportion to their body mass.

   Application question: Which of the following animals would need to eat more per gram of body mass to maintain body heat?

   A. A mouse*
   B. A dog
   C. A cow
   D. A cat

10. *Concept:* When most metal objects are heated, they increase in size.

    Application question: What assumption are you making when you run hot water over a metal lid on a glass jar to loosen the lid?

    A. Both the glass jar and metal lid increase in size in the same proportion.
    B. The glass increases in size in a greater proportion than the metal lid.
    C. The metal lid increases in size in a greater proportion than the glass jar.*
    D. Glass and metal do not stick together as much in water.

11. *Concept:* When most metal objects are heated, they increase in size.

    Application question: If a nickel with a hole in it is heated, what will happen to the size of the hole?

    A. The hole will decrease in size.
    B. The hole will stay the same size.
    C. The hole will increase in size.*
    D. The hole will become irregular.

12. *Concept:* Steam can be at the same temperature as boiling water, but steam has the energy of vaporization that is released when it condenses to water.

    Application question: Which of the following burns will tend to be most damaging to the skin?

    A. Burns caused by boiling water
    B. Burns caused by steam*
    C. Burns caused by water vapor
    D. Burns caused by hot tap water

13. *Concept:* The color of an object is determined by the wavelengths of light that it reflects and those it absorbs.

    Application question: Which of the following best explains why most plants are green?

    A. Green light is reflected from the plants.*
    B. Green light is absorbed by the plants.
    C. Green light is not needed by the plants.
    D. Green light is harmful to plants.

EXAMPLES FOR THE CREATIVITY DOMAIN

Fostering Creativity Using Creative Situations (Creativity Domain)

The use of creative writing is a way to foster creativity in students. Conceptual understanding can be monitored with this type of assessment as well. The sample questions in this section can be used to stimulate the imagination and incorporate as many concepts as the teacher deems necessary.

The teacher should set the stage for the students and could introduce the writing activity like this:

> I am going to ask some questions (or describe some situations) that will give you a chance to see how you are at thinking up new ideas and solving problems. I want you to use all the imagination and thinking ability you have. This is like a game to exercise your brain. You will have a chance to use your imagination in thinking up ideas and putting them into words. Try to think of as many ideas as possible. Try to think of interesting, unusual, and clever ideas. It does not matter if you have the same ideas as others in the class or if you suggest something that no one else thinks up.

Here are two sample questions:

Elementary: "What would the world be like if it rained chocolate drops?"

Secondary: "What would happen if lakes became frozen from the bottom?"

Situation statements should be related to the unit of instruction so that the students can connect the assessments and what they are studying. Care should be taken, however, to ensure that the unit of instruction does not center on the situation statement to such an extent that this measure of creativity becomes a test of knowledge.

Teachers may want to present situations related to concepts being studied and have students write their responses. Sample situation statements could include the following:

First grade: "Bobby woke up and found dinosaurs in his yard."
What would your reactions be?

Third grade: "Suppose we lived in a world without insects."
What would your reactions be?

Fifth grade: "Pretend that there was no more pollution."
What would your reactions be?

Seventh grade: "Suppose there was no more disease in the world."
What would your reactions be?

General: "Jane stopped at the gas station to buy gas for her car."
What would your reactions be?

Scoring This Creativity Task

The rationale behind this measure is to provide a thought-provoking situation appropriate to the ability and experiences of the students to be assessed and to have students write as many pertinent and imaginative responses to the situation as possible. The number of such responses will provide a clue to their overall creativity. These questions can be used to look at creativity by examining two factors: (a) the number of questions asked and statements made by the student and (b) the quality and uniqueness of those questions and statements.

Care must be taken to correctly frame the interpretation of student responses. Suppose the students were asked to respond to the situation statement, "Chris woke up and found dinosaurs in his front yard." A student could say that this occurred because the earth went through a time warp. This could certainly be considered unique, but if many students said that, the response would simply be pertinent. A cartoon show may have been a common experience that generated this thinking among numerous students. You could only make the determination of frequency by examining all of the student responses.

Student responses could be evaluated by using a coding system based on relevance, pertinence, and uniqueness. Consider the following examples:

| **Statement:** | Suppose you got up one morning and found that there was no gravity. |
| **Sample student response:** | Dogs would chase cats. |

This response would be judged irrelevant because it is not related to the question. If another student responded that we would float away, the response would be pertinent but not particularly creative. A unique response might be that he or she would be able to jump very high to pick fruit from trees. After evaluating student responses as irrelevant, pertinent, or unique, the responses in each category should be counted.

A Reminder: Foster Creativity, Don't Stamp It Out

The foregoing kind of activity can be conducted from time to time for mental stretching. The familiarity of the topic statement will very likely affect the range of responses. Validity and reliability are major concerns when using this kind of assessment for making judgments about student creativity. These can be fun activities to stretch thinking, but at no time should students be labeled on the basis of the results in this kind of assessment activity. The creative spirit should be fostered and not dampened.

Encouraging Analogical and Metaphorical Thinking (Creativity Domain)

Encourage analogical and metaphorical thinking in the classroom by presenting these opportunities to students. Students should have opportunities to generate analogies and then explain what the analogy supports and what it does not support.

As an example, students could be given analogies about human biology. Students could be asked to first complete correspondences as follows:

The heart is like _____.

The brain is like _____.

The skeleton is like _____.

Students could then be asked to further analyze the validity of the analogy. For the phrase "The brain is like_____," a student response might be, "The brain is like a computer. This comparison works because parts of the brain process information. This comparison is not quite accurate because our brains can think independently, but computers do not really think."

The assessment of these kinds of activities is not a particularly easy task because of the subjectivity involved. This may be the kind of activity on which students are asked to reflect over time about their capabilities. Schools want to foster students who are good thinkers, and analogical and metaphorical thinking support higher-order thinking. These are obviously not easy attributes to assess, but evidence of having these opportunities in the curriculum can be documented.

Drawing Production as a Creative Thinking Assessment (Creativity Domain)

Creativity in science may be ignored in the traditional science classroom, but the creative spirit is often an attribute of scientists. Students can be exposed to situations in which they are expected to propose visual solutions, as in the following example. Students might be presented with this scenario:

A scientist was making a drawing and was interrupted before finishing it. You found the paper and are going to help finish the drawing. You may complete the drawing in any way that you like. The scientist will be back in 15 minutes, so your work should be finished in that time.

How do you assess this drawing production (DP)?

DP allows students of most ages and ability groups to interpret and complete what they conceive to be significant for the development of a creative product. This approach has been viewed as an important addition to the culture-fair assessment of creative potential (Jellen & Urban, 1986). The Test for Creative Thinking-Drawing Production (TCT-DP) was developed in an attempt to use a more holistic and gestalt-oriented approach to examining creativity (Urban, 2004).

The DP construct is also supported by those components of creative thought that can be found throughout the existing literature on creativity and creativity testing. These components are fluency, flexibility, originality, and elaboration. Three other components of creativity that can also be assessed via this approach are risk taking, composition, and humor as a cognitive-affective ability capable of freeing the mind from concrete or unpleasant realities (Jellen & Urban, 1986; Urban, 2004). Students are presented with figural elements or fragments and given the task of completing the drawing. Fourteen key elements serve as evaluation criteria for the TCT-DP. Information about the TCT-DP can be accessed online (http://www.tvtc.com).

Activities to Foster Creativity

Creativity is not easy to assess, but creativity can be fostered in a classroom environment in which students are engaged in activities that include projects and scenarios to which they respond. Examples, such as the Floodville sample in Figure 6.6, can be linked with natural disasters, environmental issues, ecosystems, water recreation, water pollution, city planning, economics, mapping skills, governmental involvement, and numerous real-world events.

A sample like Floodville might begin with a scenario created by the teacher or by the students and the teacher. To help in creating this scenario, the newspapers or other media news could serve as a source for details of real-world events, and a scenario that has local relevance can be generated. The local weather history of an area could also be brought into the scenario. Scenario development can be accomplished through four major steps. The teacher may want to model this process with a teacher-generated scenario and drawing, and then students, working in cooperative groups, could also generate their own scenarios.

Floodville (Creativity Domain)

1. *Generate a scenario:* Floodville is a town that was established on the banks of the Floody River, which is deep enough to allow barge travel and support recreational activities such as boating and fishing. In fact, many people own homes on riverfront property. Floodville sits in a valley surrounded by hills that were once covered with trees. The trees were a source of valuable timber, and the hills are low enough for farming. Floody River floods in the spring but usually causes little or no major property damage and only a few road closings. About every 10 years, a major flood occurs and causes several million dollars worth of property damage.

Figure 6.6   Map of Floodville and Floody River

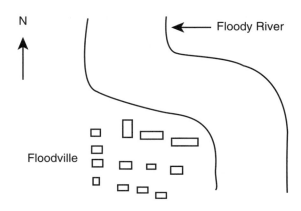

2. *Brainstorm some questions or issues for student reactions:* The government proposes to have Floodville relocate to a higher elevation. This would involve moving 250 families, a school, two churches, two small factories, a grocery store, the city offices, and two gas stations. Imagine that you are a Floodville resident: What are some kinds of information you would want to have before you would support such a decision? What would you propose as alternatives to relocation? What rebuttals could be made to support not relocating the town? What changes to the surrounding environment likely worsened the flooding? How might runoff from farms create problems? Would a dam or flood wall help?

3. *Relate to pertinent issues:* This kind of scenario can also be moved to issues such as whether building should be permitted on flood plains (or coastal areas subject to hurricanes or in earthquake-prone areas). Role-playing is also a viable option for addressing these issues.

4. *Foster creativity:* What constitutes creativity in solutions to scenarios depends on the novelty of solutions presented, and this novelty should likely be grounded in the reality of a workable solution. A suggestion is to focus on assessing problem solving and higher-order thinking skills. Marzano, Pickering, and McTighe (1993), in their dimensions of learning model, provide a set of rubrics on which to assess these skills.

ASSESSING ATTITUDES, PREFERENCES, AND PROCESSES

The three instruments that follow (Figures 6.7–6.9), taken from various sources, are designed to assess aspects of students not readily observed or accessible through curriculum content.

Figure 6.7 Attitudes About Science, Technology, and Society (STS) (Attitude Domain)

Circle the letter(s) that is/are the best indication of your feelings about the statement.

SA = Strongly Agree
A = Agree
N = Neutral
D = Disagree
SD = Strongly Disagree

	Strongly Agree	Agree	Neutral	Disagree	Strongly Disagree
1. Technology will eventually solve all societal problems.	SA	A	N	D	SD
2. Problems resulting from science or technology hardly ever affect me.	SA	A	N	D	SD
3. I would like to know more about how to reduce pollution.	SA	A	N	D	SD
4. I would like to know more about alternative energy sources.	SA	A	N	D	SD
5. I would like to know more about genetically modified foods.	SA	A	N	D	SD
6. Clearing rain forests for farming affects only the environment in areas where the forests were.	SA	A	N	D	SD
7. I want to know how STS issues affect me.	SA	A	N	D	SD
8. If I knew more about STS issues, I could do more about them.	SA	A	N	D	SD
9. Most people will not act on STS issues even if they understand why action is needed.	SA	A	N	D	SD
10. I know all I care to know about environmental problems.	SA	A	N	D	SD
11. Consumers like me are really trying to solve STS problems.	SA	A	N	D	SD
12. The government is really trying to solve STS problems.	SA	A	N	D	SD
13. All science classes should include STS issues and topics in the curriculum.	SA	A	N	D	SD
14. I would be willing to pay at least 10% more to buy something if the manufacturers used this money to reduce their pollution.	SA	A	N	D	SD

	Strongly Agree	Agree	Neutral	Disagree	Strongly Disagree
15. Members in a society have the responsibility to develop an appreciation and respect for the rights of others within the society.	SA	A	N	D	SD
16. STS education should use a problem-solving approach.	SA	A	N	D	SD
17. Declining environmental quality poses a serious threat to the future health and well-being of most Americans.	SA	A	N	D	SD
18. Open discussions, like town hall meetings, should be a main strategy used in STS teaching.	SA	A	N	D	SD

What is the number one STS-related issue facing today's world? Why do you think this?

Figure 6.8   Assessing Attitudes and Preferences in Science, Grades 4 Through 12 (Attitude Domain)

For statements 1 through 18, circle the number that best indicates your view.

Use the following rating scale:

4 = Almost Always

3 = Sometimes

2 = Seldom

1 = Never

	Almost Always	Sometimes	Seldom	Never
1. Science classes are fun.	4	3	2	1
2. Science classes increase my curiosity.	4	3	2	1
3. What we study in science classes is useful to me in daily living.	4	3	2	1
4. Science classes help me test ideas I have.	4	3	2	1
5. My science teacher admits to not having answers to some of my questions.	4	3	2	1
6. Science classes help me with skills I use outside of school.	4	3	2	1
7. My science class deals with the information produced by scientists.	4	3	2	1
8. Science classes are exciting.	4	3	2	1
9. Science classes provide a chance for me to follow up on questions I have.	4	3	2	1
10. My science teacher encourages me to ask questions.	4	3	2	1
11. All people can do and practice basic science.	4	3	2	1
12. Being a scientist would be fun.	4	3	2	1
13. Being a scientist would make a person feel important.	4	3	2	1
14. Science classes are boring.	4	3	2	1
15. Being a scientist would be lonely.	4	3	2	1
16. Being a scientist would make a person rich.	4	3	2	1
17. Being a scientist would mean giving up some things of interest.	4	3	2	1
18. Scientists discover information that is difficult to understand.	4	3	2	1

For the subjects listed 19 through 26, circle "1" for the subjects you like the best, and circle "2" for the subjects you do not like as well. Last, circle the name of your favorite subject.

19. Foreign Language	1	2	23. Physical Education	1	2
20. Science	1	2	24. Language Arts	1	2
21. Mathematics	1	2	25. Reading	1	2
22. Social Studies	1	2	26. Music	1	2

Source: Adapted from National Assessment of Educational Progress (NAEP). (1978). *The Third Assessment of Science (1976–1977).* Denver, CO: Author.

Figure 6.9   Science Process Test

Name: _____   **Date:** _____

Teacher: _____   **Course:** _____

Circle the letter that best applies to you.

1. Gender: a = female; b = male

   a   b

2. Ethnicity: a = Asian; b = Black; c = Hispanic, Latin; d = Caucasian; e = Native American, Alaskan Native

   a   b   c   d   e

   Other (please write in): _____

3. Grade: a = 6, 7, 8; b = 9; c = 10; d = 11; e = 12

   a   b   c   d   e

For each question, circle the letter of the best answer.

Making Observations

4. Which of the following is an observation only?

   A. The piece of metal is red so it must be hot.
   B. The street is wet so it must have rained.
   C. The table looks like it is made of wood.
   D. The block is orange.

5. Which of the following could be observed with the sense of sight?

   A. A change in the temperature of the air
   B. A change in the height of plants
   C. A change in the sweetness of a new chemical
   D. A change in the noise made by an engine

Using Space-Time Relationships

6. If runners A and B start at the same time and they arrive at the finish line (C) at the same moment, who ran faster?

   A. A ran faster than B.
   B. B ran faster than A.
   C. A and B ran at the same speed.
   D. B ran slower than A.

7. Which shape of shadow could not be formed by a solid cylinder?

   A. A circle
   B. A square
   C. A rectangle
   D. A triangle

(Continued)

Figure 6.9 (Continued)

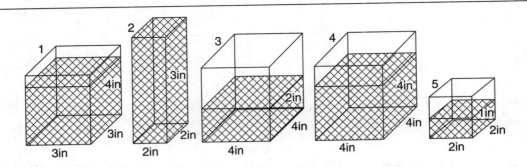

8. According to the preceding diagram, which two containers have approximately equal volumes of water in them? (Drawings are not to scale.)

   A. Containers 1 and 2
   B. Containers 2 and 3
   C. Containers 3 and 5
   D. Containers 2 and 5

Classification

9. The following table has some background information about students in Jones Elementary School.

1. Name	2. Gender	3. Birthday	4. Nationality	5. Year Entered School
M. John	Female	June 1998	American	2004
B. Thames	Male	March 1998	British	2004
A. Shirley	Male	December 1997	American	2004
R. Thompson	Female	May 1998	American	2004
R. Ali	Male	October 1997	Indonesian	2004
B. Ghombal	Male	August 1997	Portuguese	2004

   If you wanted to sort these students into at least two different groups, which of the following categories could *not* be used to do so?

   A. Gender (male or female)
   B. Year of birth
   C. Nationality
   D. Year entered school

10. Which would be the best feature to use in classifying the following shapes into two groups?

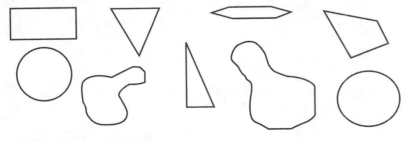

   A. Square versus not square
   B. No straight sides versus four straight sides
   C. Circle versus triangle
   D. Curved edge versus straight edge
   E. Odd number of sides versus even number of sides

11. The hotter the water, the faster sugar will dissolve. Look at the jars in the following drawing. Each jar has the same amount of sugar. Put the jars in order from the slowest rate for sugar to dissolve to the fastest rate to dissolve.

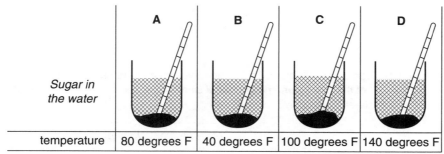

	A	B	C	D
temperature	80 degrees F	40 degrees F	100 degrees F	140 degrees F

A. A, B, C, D
B. B, A, C, D
C. C, B, D, A
D. D, C, B, A

Using Numbers

12. Which of the following groups of objects presents the objects in order, from left to right, of the smallest to the largest number?

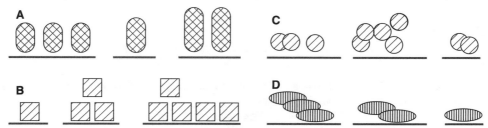

13. What is the next number in this number sequence?

2   3   5   8   12   17   _?_

A. 19
B. 23
C. 24
D. 28

14. Yesterday it was –5°C. Today it is 10°C. How many degrees warmer is it today than it was yesterday?

A. 5°C warmer
B. 10°C warmer
C. 15°C warmer
D. –5°C warmer

Measuring

15. Normal human body temperature is about 37°C. The body temperature of someone who is ill ranges from about 36°C to 42°C. Which thermometer would be the best to use for measuring any human body temperature?

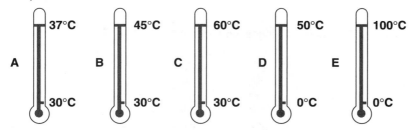

Figure 6.9 (Continued)

A. A
B. B
C. C
D. D
E. E

16. Four students were each given their own plant. To practice their measuring skills, each student was asked to measure the height of the plant four times during a single class period. Which student do you think measured most carefully and precisely?

Student	Measurement Number			
	1	2	3	4
Rusty's plant heights	3 cm	6 cm	10 cm	8 cm
Mike's plant heights	4 cm	5 cm	5 cm	4 cm
Karen's plant heights	2 cm	10 cm	4 cm	8 cm
Carol's plant heights	8 cm	3 cm	2 cm	1 cm

A. Rusty
B. Mike
C. Karen
D. Carol

17. Dan and Tom each shot 20 foul shots on each day for 5 days. Using the following graph, on how many days did Dan make more baskets than Tom?

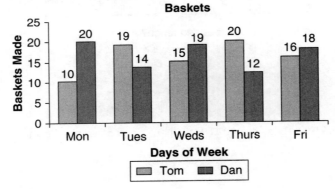

A. 1 day
B. 2 days
C. 3 days
D. 4 days
E. 5 days

Communicating

18. What object has 6 equal faces, 8 corners, 12 edges, and volume?

A. A cube
B. A square
C. A sphere
D. A cone
E. A hexagon

19. A tennis ball was dropped from several different heights, and the height the ball bounced was recorded each time the ball was dropped. Which of the following would be the best method to report the data collected?

    A. A written paragraph
    B. A tally of the number of bounces
    C. A frequency distribution
    D. A bar graph
    E. A pie chart

20. Mary wants to make a diagram of the school classroom on a piece of notebook paper. Which of the following choices would be a convenient scale for her to use?

    A. 1 inch = 1 mile
    B. 1 inch = 1 centimeter
    C. 1 inch = 1 yard
    D. 1 inch = 1 acre
    E. 1 inch = 1 inch

Making Inferences

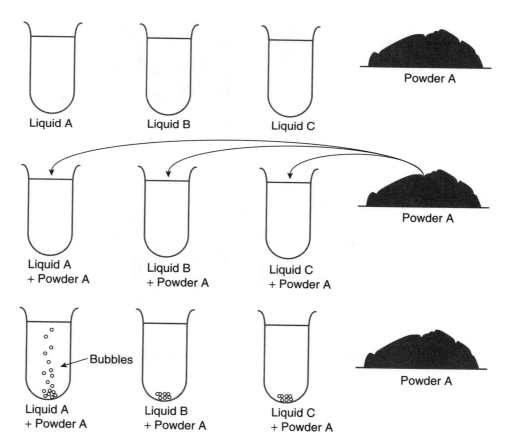

21. Which inference is best supported by the above diagram?

    A. Liquids A and C are the same.
    B. Liquids A and B are not the same.
    C. Liquids B and C are both the same.
    D. Liquids A, B, and C are all the same.

(Continued)

Figure 6.9 (Continued)

Predicting

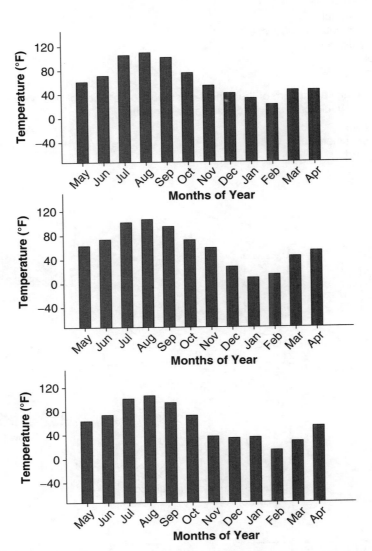

22. The average temperature recorded each month during the past 10 years is shown on these graphs. Based on this evidence, which month do you think will be coldest next year?

    A. June
    B. September
    C. November
    D. January
    E. February

23. The same amount of a gas that is less dense than air is placed in each of the following balloons. Which balloon do you think will float up the fastest?

	Weight in Pounds
Balloon A	1000
Balloon B	800
Balloon C	500
Balloon D	200

24. Look at the objects below. Which item do you think will sink fastest in a pan of water?

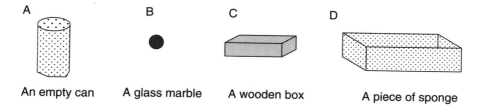

A An empty can   B A glass marble   C A wooden box   D A piece of sponge

A. An empty can
B. A glass marble
C. A wooden box
D. A piece of sponge

25. Dan and Dawn want to know if there is any difference between the mileage expected from bicycle tires from two different manufacturers. Dan will put one brand on his bike, and Dawn will put the other brand on her bike. Which of the following variables would be most important to control in this experiment?

A. The time of day the test is made
B. The number of miles traveled by each type of tire
C. The physical condition of the cyclist
D. The weather conditions
E. The weight of the bicycle used

26. A group of students conducted an experiment to determine the effect of heating on the germination (sprouting) of bean seeds. Which of the variables in the following list is least important to control in this experiment?

A. The temperature to which the seeds are heated
B. The length of time the seeds are heated
C. The type of soil used
D. The amount of moisture in the soil
E. The size of the container used for growing each seed

27. A student wants to know how the amount of acid rain affects the fish population. She takes two jars and fills each of the jars with the same amount of water. She adds 50 drops of vinegar (acid) to one jar and adds nothing extra to the other. She then puts 10 similar fish in each jar. Both groups of fish are cared for (oxygen, food, etc.) in identical fashion. After observing the behavior of the fish for a week, she makes her conclusions. Without adding another variable, what could you suggest to improve this experiment?

A. Prepare more jars with different amounts of vinegar (acid).
B. Add more fish to the two jars already in use.
C. Add more jars with different kinds of fish and different amounts of vinegar in each jar.
D. Add more vinegar to the two jars already in use.

(Continued)

Figure 6.9 (Continued)

Interpreting Data

28. The following data are taken from an experiment:

Temperature (average)	Mass of Seed (grams)	Water Uptake (ml/day)	Exposure to Light (min/day)	Plant Height (cm/20 days)
20°C	2.2	10	20	20.2
50°C	2.3	10	20	20.3
30°C	2.3	10	20	20.2
25°C	2.1	10	20	20.3
25°C	2.3	10	30	21.9
25°C	2.2	10	40	22.8
20°C	2.2	10	30	21.8
20°C	2.1	20	30	21.9
20°C	2.2	30	30	22.0

Based on the data in the table, what factor do you think has the greatest influence on the rate of plant growth?

A. The temperature where the plant is grown
B. The seed mass
C. The amount of water uptake each day
D. The length of the time the plant is exposed to the light

29. Here is an experiment that shows how much some peanut plants grew in 20 days.

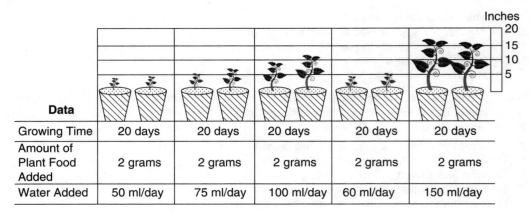

Data					
Growing Time	20 days	20 days	20 days	20 days	20 days
Amount of Plant Food Added	2 grams	2 grams	2 grams	2 grams	2 grams
Water Added	50 ml/day	75 ml/day	100 ml/day	60 ml/day	150 ml/day

Look at the information given in the graphic. Which is the best conclusion for this experiment?

A. The more plant food that was added, the faster the plant grew.
B. The more plant food that was added, the slower the plant grew.
C. The more water that was added, the faster the plant grew.
D. The more water that was added, the slower the plant grew.

Formulating a Hypothesis

30. Bob set up two identical bowls. Both contained sugar water, and both were open to the air. One was placed in the dark, whereas the other was placed in the light. What one item differs from one setup to the other?

A. The exposure to light
B. The shape of the bowl
C. The exposure to air
D. The amount of sugar

31. Which of these statements is the best example of a hypothesis?

    A. This magnet picked up 12 paper clips.
    B. The milk in this bottle froze in 20 minutes.
    C. The house plant may have died from being watered too much.
    D. The leaves on that maple tree have all turned red.
    E. At that rate, the pool filled in 10 minutes.

32. Examine the following data table, and select the most appropriate hypothesis regarding dissolving time and water temperature.

Substance	Average Dissolving Time (in seconds)			
	Water 20°C	Water 40°C	Water 50°C	Water 60°C
20 g of sugar	80	40	20	5
20 g of salt	60	30	16	3

    A. There is no difference in dissolving time of the substances because of water temperature.
    B. The lower the temperature of water, the less time needed to dissolve the substances.
    C. The higher the temperature of the water, the less time needed to dissolve the substances.
    D. It is impossible to make a hypothesis from the information given in the chart.

Defining Operationally

33. Which one of the following is written as an operational definition?

    A. Because the density of oil is less than the density of water, when water is mixed with oil, the oil will float on the surface of the water.
    B. The speed of a supersonic jet is similar to the speed of sound waves.
    C. When you drive your car at a speed of 30 miles per hour, you have to push the brake pedal 300 feet before the line or point where you are planning to stop.
    D. The speed of a car will decrease when it has to turn right or left.

Experimenting

34. A student wants to test to see if the color of cloth influences the amount of heat absorbed. He plans an experiment using two colors of cloth to wrap two different glasses, each containing the same amount of water. One glass is wrapped with green cloth, and the other is wrapped with yellow. He places the glasses in a sunny spot and sets a thermometer in each glass to observe the temperature. What things can you suggest to improve his testing?

    A. Add to the number of glasses covered with the cloth.
    B. Reduce the amount of water in each glass.
    C. Prepare more containers, each covered with a different color of cloth.
    D. Double the size of the cloth used to cover the glass.

35. Eight bean seeds were allowed to germinate and were then divided into four groups of two seeds each. One group was grown under red light, another under green light, another under blue light, and the fourth under white light. At the end of 2 weeks, the growth of each group of plants was measured to see which group of plants had grown the most. This experiment could best be improved by doing which of the following?

    A. Giving more water to the plants grown under the red light
    B. Increasing the number of seeds grown in each of the four groups
    C. Growing just the plants under white light in sandy soil, but growing all others in humus soil
    D. Adding one more group of two seeds to the experiment and growing them under purple light

Figure 6.9 (Continued)

36. Gloria wants to determine the temperature best suited for fish. Which of the following procedures is the best to use to determine this?

   A. Get 5 aquaria (fish tanks) and place 5 similar fish in each aquarium. Keep the temperature in each aquarium constant at 25°C.

   B. Place 5 fish in one aquarium. At intervals of 10 minutes, change the water temperature from 10°C to 15°C, to 20°C, to 25°C, and, finally, to 30°C. Observe the behavior of the fish after each change in temperature.

   C. Get 5 aquaria and place 5 similar fish in each aquarium. Keep the temperature of the water constant at about 25°C, and observe the behavior of the fish in each aquarium.

   D. Get 5 aquaria, and place 5 similar fish in each aquarium. Each aquarium should have one of these five water temperatures: 15°, 20°, 25°, 30°, or 35°C. Observe the behavior of the fish in each aquarium.

Source: Adapted from the Mason City Schools, Mason City, Iowa, 1993; and from National Assessment of Education Progress (NAEP). (1978). *The Third Assessment of Science (1976–1977).* Denver, CO: Author.

7

Assessment Examples for Grades K Through 4

In this chapter, you can find suggestions for the following:

- Structuring a fair test
- Asking concept-related application questions
- Assessing student attitudes toward science
- Student self-assessments
- Assessing students' perceptions of scientists

APPLYING PROCESS SKILLS AND EXPERIMENTAL DESIGN

Does the Brand of Bubble Gum Make a Difference in Bubble Size?

The use of process skills can be supported by framing experiences in a context to which children can relate. For example, if students wanted to know which brand of bubble gum could be used to blow the biggest bubbles, they might first ask the obvious question, "Which brand of bubble gum makes the biggest bubbles?" The second question asked might be, "How could we study this question?" From here, the students and the teacher could set up a plan of action to try to find out.

In setting up a plan of action to address the question of interest, class discussion could be guided toward conducting a fair test. Students might brainstorm a list of questions that would need to be considered, such as,

- How many brands of gum should we test?
- Does sugar make a difference?
- How long should we chew the gum before blowing a bubble?
- What styles or techniques of bubble blowing should we use?

With younger students, the idea of controlling variables may be too abstract, but a discussion of why they would want to try to keep things the same when doing a fair test is appropriate. The students could design a plan together in which they would all use the same brand of sugarless gum. They could decide on how long to chew the gum and then try a technique to blow a bubble. Students could be asked to estimate how big of a bubble they thought they could blow. Students could be asked to really focus on the techniques that they used and then write descriptions of their techniques.

Students could then go on to test two brands of sugarless gum. This could be done as a cooperative group project where one student could be the bubble blower, one the data collector, one the recorder, and one the quality control agent who makes sure that the plan is followed. After students have collected their data, they could be asked to average their data and write about their findings. All of the class data could be displayed, discussed, and summarized. Students could also write about their findings and develop a report to send to the bubble gum manufacturer.

Testing the water-absorbing capabilities or strengths of various paper towel brands could serve as an alternative to the bubble gum fair test. In addition, flying paper planes based on certain modifications could offer an inexpensive alternative. Blowing soap bubbles from various concentrations of water, glycerin, and liquid soaps could also be an opportunity for setting up an investigation.

A checklist approach, as set in the following table, could be used to document class progress in process skill use and in designing experiments.

			Improvements Needed
Yes	No	Students were able to generate a question to investigate.	
Yes	No	Students were able to generate a plan to follow.	
Yes	No	Students were able to set up a fair test.	
Yes	No	Students were able to set up a data table.	
Yes	No	Students were able to conduct a fair test.	
Yes	No	Students were able to record data.	

			Improvements Needed
Yes	No	Students used data-gathering tools in appropriate ways.	
Yes	No	Students were able to tabulate their data.	
Yes	No	Students were able to graph their data.	
Yes	No	Students were able to develop a conclusion and report their findings.	

APPLICATION ASSESSMENT ITEMS FOR GRADES K THROUGH 3

Figure 7.1 presents some sample assessment items for the application domain, appropriate for Grades K through 3.

Figure 7.1 Sample Application Assessment Items: K–3 (Application Domain)

1. *Concept:* Birds hatch from eggs.

   Application question: You have found some eggs in a nest in a tree. You watch the nest and the eggs hatch. Draw a picture of the kind of animal that you think will come from the eggs. (A picture could be used to elicit student responses.)

2. *Concept:* The cyclic movement of water between the land and the air is called the water cycle.

   Application question: Yesterday, while playing outside, you made a mud pie. The pie was very runny, so you left it outside. If tomorrow is sunny, draw a picture of what the mud pie will look like when you go out to play with it.

3. *Concept:* When resistance is increased, it takes more force to move an object.

   Application question: Which child is working harder to move the object?

   (Two similar pictures are needed. One should be of a child pulling an empty sled or wagon, and the second picture should show an object placed on the sled or in the wagon.)

4. *Concept:* Environments and habitats are all changing. Some changes are a result of natural forces, and some are a result of the actions of people.

   Application question: A company is going to build a road on some land with forest. Draw one picture of what you think the land looked like before the road was built. Then draw a picture of how the area looked when the road was being built.

5. *Concept:* Meters are used for measuring distance.

   Application question: The teacher has asked you to measure the length of your classroom. What unit of measurement will you use? Circle the best answer. Tell why you think this answer is best.

   Kilometers *Meters* *Grams* *Liters*

ASSESSING ATTITUDES ABOUT SCIENCE

Figures 7.2 and 7.3 present ways to elicit young students' attitudes about science. Figure 7.2 uses visual cues, and Figure 7.3 is to be used in conjunction with interviewing.

Figure 7.2 How I Feel About Science: Grades K–3

Name: _____ **Grade level:** _____

How do you feel about your science class?

After each sentence, place an X on the face that shows how you feel.

	Yes	*It's OK*	*No*
1. Science time is fun.	☺	😐	☹
2. Things I learn in science help me understand things at home.	☺	😐	☹
3. I can do science things now.	☺	😐	☹
4. We do fun things in science class.	☺	😐	☹
5. In science, my teacher likes to have me ask questions.	☺	😐	☹
6. I might want to have a science job when I grow up.	☺	😐	☹

Students could also be asked to draw a picture of themselves doing something science related.

Figure 7.3   How I Feel About Science: Grades 2–4

What are some of your thoughts about science?

1. What would you write or say to finish this sentence? To me, science is . . .

2. Tell all the ways you use science in a day.

3. What are some things you like about science?

4. What are some things you do not like about science?

5. If someone told you that science is not important, tell why you would agree or disagree with them.

6. What kinds of things do you think scientists do?

STUDENT SELF-ASSESSMENT

After completing an activity on measuring various objects, students can indicate what they were able to do on an instrument such as the one shown in Figure 7.4; they should have their work available so they can share what they have done. Students could be asked to draw a star by the things that they can do. This is a way to engage students in self-assessment at an early age. The *National Science Education Standards* (NSES) (National Research Council [NRC], 1996) recommend that students use conventional measuring instruments, such as rulers, thermometers, and balances; the student checklist could be modified according to the instruments used. In science, measurements are usually metric based, but with younger students, the use of the English system may be appropriate as their initial frame of reference. The activity selected for practice with measuring should be developmentally appropriate for students. For example, if students are just learning to measure with a ruler, the ruler scaling should also be appropriate for these students.

Figure 7.4   I Can Measure With a Ruler! (Integrates Domains)

1. I can start from the left end of the ruler.

2. I can measure in inches.

3. I can measure in feet.

4. I can measure four (4) things.

5. I can draw a picture of each thing I measured.

6. I can write the measurement under each picture.

Performance Activity

To accompany these "I can" statements, stations could be set up where students are asked to make measurements. Lines of various lengths could be drawn on a piece of paper, and students could measure and record the lengths of the lines. Students could also be asked to cut various lengths of yarn or string. Students could measure various objects in the room: desk tops, books, pencils, and so on. Pieces of masking tape could be placed on the wall so students could mark and measure their heights. The masking tape could be marked with pre-measured feet and inches so that the task is easier for students. Paper with one-inch grids could be printed, and students could construct paper measuring tapes of various lengths and use these to make longer measurements.

If students are at an age where they are learning metric measurements, this could also be incorporated. Many classrooms also have manipulatives that could be incorporated in this type of activity. If students are studying fractional parts, this kind of measuring activity could be used for engagement and exploration.

ASSESSING PERCEPTIONS ABOUT SCIENTISTS

Students may have stereotypic views of scientists and what scientists do. The following assessment could be used at the beginning of the year and again at the end of the year to see if student perceptions have changed. For students' perceptions to change, it is likely that classroom instruction would need to address misperceptions about scientists. It is possible to get a look at students' perceptions by having them draw pictures of their ideas about scientists.

An "if-then" to consider: If children hold stereotypic views of scientists and what scientists do, then what will you, as the teacher, do to change their perceptions? Fort and Varney (1989) reported on a study in which students were asked to portray scientists in words and pictures; they found that based on the data, students mostly viewed scientists as male, white, and benevolent. Barnum (1996, 1997) invited science teachers to participate in a study in which their students drew scientists and then wrote about them. Barnum also suggested asking students to draw themselves doing science and recommended conducting taped interviews with the students.

Draw a Scientist (Nature of Science Domain)

Purpose: To gain insight into how children view scientists and what they think scientists do.

Preassessment: Ask students to draw a picture of a scientist. Also ask them to describe what they have drawn and to tell what they think a scientist does.

Interim instruction: If students hold stereotypic views of scientists, then science instruction must include experiences that foster changes in student thinking about scientists. Teachers could work to do this by recognizing opportunities that provide evidence that scientists are people and that doing science is a human endeavor. Ideally, raising student awareness of the diversity that exists among scientists should be an ongoing effort in the classroom.

Postassessment: Ask students to draw another picture of a scientist. Also ask them to describe what they have drawn and to tell what they think a scientist does. This postassessment could be used at any time, but it may be most useful at midyear or at the end of the year.

Evaluation of the drawings: Students could be asked to describe any differences between the preassessment and postassessment pictures. If their perceptions have changed, they could also be asked to give reasons for the changes. Barnum (1996, 1997) and Fort and Varney (1989) offered guidelines for evaluating students' drawings. More information about the Draw-a-Scientist Test (DAST) and interpreting drawings can be accessed online. Thomas, Pedersen, and Finson (2001) designed DASTT-C, a version of DAST where students are asked to draw a science teacher.

8

Assessment Examples for Grades 5 Through 8

In this chapter, you can find suggestions for the following:

- Structuring performance tasks
- Using a context to anchor multiple-choice items
- Incorporating electronic or video portfolios
- Involving parents in portfolio review
- Structuring peer assessment for cooperative groups
- Structuring group assessment of cooperative roles
- Structuring class assessment of group project presentations
- Using student self-assessment of role in a group project

DEVELOPING A GROUP PERFORMANCE TASK

Evolution of Toys: Back to the Future (Integrates Domains)

Developing and crafting a performance task is, in itself, an evolutionary process. This performance task core idea could be extended in numerous ways to incorporate a variety of skills and disciplines, and as a rule, good instructional tasks have the potential to become good performance tasks.

Performance Task Core Idea

A major toy company has selected the theme "Evolution of Toys: Back to the Future." This toy company has sent out a request for toy designs, and your team of four classmates is interested in designing and submitting a toy. Before your team actually designs a toy, you want to know what toys were like years ago. You can decide upon the time frame. What were toys like before the advent of electronics? Electronic toys and games have changed in the time since their advent. Using your experiences with toys and other resources, you plan to first compare the toys of today with the toys from years ago. Let's say you decided to look at the toys of the 1940s.

The team project description might go something like this. Based on the technology used in the toys children played with in the 1940s, describe three ways that the life of a child of the 1940s and your life in the 2000s vary. Your team should prepare a two- to three-page written paper that describes the two different toys on which you base your inferences about the life of a child in the 1940s. Include a drawing or picture of each of these toys and tell how the toy worked. Your group will also be expected to give a 5-minute oral presentation about your findings to the rest of the class.

Performance Expectations

The following are possible performance expectations:

- Based on the technology used in the toys, inferences are made about how a child's life in the 1940s differs from a child's life today.
- Inferences are supported by a variety of references and resources.
- References and resources are documented in a bibliography.
- A two- to three-page word-processed paper is developed.
- Drawings of toys and descriptions of how the toys work are included.
- A 5-minute oral presentation is to be made by the group.
- Each member of the group is expected to contribute to the presentation.
- In group work, intrapersonal skills are evaluated.
- Unique approaches and creativity in products are noted.

Possible Extensions of the Core Task

- The toys could be drawn to scale.
- Scale drawings could be completed using a computer platform.
- Models of toys could be constructed.
- Production costs for building the toys could be calculated.
- Marketing brochures could be designed.
- The toys could be used as the context for physics concepts.
- The history of the 1940s could be addressed.
- Creative elements could be examined.
- Students could design a new toy.

An Individual or Group Performance Task: Taking Care of Garbage (Integrates Domains)

Performance Task Core Idea

You have watched the garbage cans being picked up in your neighborhood. Knowing hundreds of garbage cans must be lifted each day, you intend to make life easier for the workers by designing a mechanical device that will lift the garbage can and empty the contents into the garbage truck. Your task is to design and make a working cardboard model of your device and then demonstrate and explain its operation as a machine.

Performance Expectations

The following are possible performance expectations:

- A scale drawing should be completed.
- A model should be constructed.
- A demonstration of the model should be given.
- The physics and engineering concepts of the model could be explained.
- Creative elements could be examined.

Possible Extensions for the Core Task

- References and resources could be documented in a bibliography.
- Connections to the history of technological design and the Industrial Revolution could be made.
- A marketing brochure could be developed.
- The model could be evaluated by an engineer.
- An investigation of the patent process could be completed.
- Famous inventors could be incorporated.
- The creative process could be addressed.

Other Ideas

Whatever the performance task, be certain to begin with a purpose for having students complete the work. Students could write and illustrate books for lower grade levels. Hot-air balloons, kite designs, mousetrap cars, digital storytelling, poetry, game designs, robotics, or countless other tasks are possible. Performance tasks could also involve projects that have service components.

PREPARATION FOR STANDARDIZED TESTS

Preparing students for standardized tests can include practice with a context followed by a selection of multiple-choice questions. This type of format—a contextual setting accompanied by multiple-choice questions—tends to support questions with greater cognitive demands. The following example, adapted by Peter Veronesi from the General Educational Development [GED] Test (1990), presents background information along with a table containing the essential facts and pertinent multiple-choice questions. An asterisk marks the best answer for each question.

Multiple-Choice Application Example:
Concepts of ABO Blood Types (Application Domain)

Contextual Background for ABO Blood Types

Blood transfusions were practiced in the 1800s, and blood was transferred from one person to another. Sometimes these transfusions were miraculously successful, whereas at other times, persons experienced pain, tingling sensations, and often, death. In 1901, the scientific basis for successful blood transfusions was identified, and it was recognized that only certain blood types could be successfully transfused to a given individual.

If a person is given a transfusion of an incompatible blood type, his or her blood cells will *agglutinate*, meaning clump together. Agglutination is due to a reaction between antigens and antibodies in the blood. An *antigen* refers to any substance that produces an immune response, and usually, the body reacts by producing antibodies. The antibodies attack the antigens, which are recognized as foreign substances.

The ABO system is one system used to categorize blood-cell antigens. The surfaces of red blood cells of types A, B, and AB contain certain substances that act like antigens. The safest transfusion of blood takes place when the donor and the recipient have the same type of blood. The most critical factors are the antigens present in the donor's blood and the antibodies present in the recipient's blood plasma. If the red blood cells containing antigens are transferred into a person whose blood plasma contains antibodies against them, clumping of the blood occurs.

Use this information to answer Questions 1 through 5:

Blood Type	Red Blood Cells Have	Plasma Carries
A	A antigen	Antibodies against B
B	B antigen	Antibodies against A
AB	Both A and B antigen	No antibodies against A or B
0	No A or B antigen	Antibodies against A and B

1. A blood bank had five problems with a laboratory technologist's work performance during the first year. Of the problems listed below, which would the technician's supervisor likely consider the most serious?

   A. Coming to work late
   B. Spilling five units of blood
   C. Working too slowly
   D. Mislabeling four units of blood*
   E. Breaking the lens on a microscope

2. From the information presented, the most probable reason some people died in the 1800s after receiving blood transfusions, whereas others recovered, was due to

   A. the type of illness the person had.
   B. the skill of the physician.
   C. how sterile the equipment was.
   D. whether or not the person received compatible blood.*
   E. how much blood was lost during the operation.

3. Suppose that each of the persons below received a blood transfusion:

   A. A four-year-old child
   B. A middle-aged man
   C. An elderly woman

In which of these people would agglutination probably occur if given the wrong blood type?

A. A only
B. B only
C. C only
D. B and C only
E. A, B, and C*

4. Which of the following would most likely result in agglutination?

A. The mixing of similar blood types
B. The separation of blood cells from plasma
C. The reaction of antigens to other antigens
D. The reaction of antibodies to foreign antigens*
E. The donation of blood to a blood bank

5. A universal receiver can receive a blood transfusion from persons of any blood type. Which blood type(s) would a person need to have to be a universal receiver?

A. Type A only
B. Type B only
C. Type AB only*
D. Type O only
E. Either type AB or O

Note: Rh factor compatibility must also be addressed in blood transfusions.

ELECTRONIC PORTFOLIOS FOR STUDENTS

A teacher may wish to have students document their work and class-related activities in an electronic portfolio. The technology exists that can make this effort very student driven. With digital cameras and the ease of use of software, students can create multimedia portfolios that document their work and, at the same time, expand their skills in technology use. Throughout the school year, a visual record of projects, presentations, and group work could be created; this would require access to equipment and software to support the effort. Software to support electronic portfolios is available, and the generation of student Web pages is also a possibility; students today tend to be technologically savvy.

The student, teacher, administrator, and parents could view the videos, electronic portfolios, or Web pages periodically during the year to look at student growth. These records could be used to provide feedback and could serve as an excellent record of each student's year. School policies should be followed with regard to confidentiality and anonymity. If a digital format was used, a review form could be sent home with the material or URL so that when parents and students view the material together, they can discuss and critique the work. Always check with school policies to be certain that Web-based materials are secure and that any consent policies have been followed. Also, you may need to make some modifications if parents or guardians have limited access to technology. Figure 8.1 gives an example of a communication that could be sent home.

Figure 8.1   Letter for Review of Student Work Sample

Dear _____,
　　　　　　　　(parent/guardian)

_____ (student's name) has created this work from our science class and would like you to review this with him or her. While reviewing the work with your child, please discuss and comment on the areas mentioned in this letter. Look for strengths in the work sample, and also look for areas in need of improvement.

Interest in work:

What did you find most interesting about your child's work sample?

Work quality:

After reviewing the work, what did you notice that was done well?

Was there something that needed a little more work?

Feedback to your child:

Write a positive comment to your child about the work sample you reviewed.

Thank you for your time and interest in your child's education. We are very proud of the work we are doing and wanted to invite you to share some of our experiences.

　　　　　　　　　　　　　　　　　　　Sincerely,

　　　　　　　　　　　　　　　　　　　_____
　　　　　　　　　　　　　　　　　　　　　　　(Student's signature)

　　　　　　　　　　　　　　　　　　　_____
　　　　　　　　　　　　　　　　　　　　　　　(Teacher's signature)

STRUCTURING PEER ASSESSMENT FOR COOPERATIVE GROUPS

Cooperative groups are frequently used for work in classes, and peer assessment can help group members assume more personal responsibility for actions within the group. This assessment can include both assessment of each group member by their peers and an assessment of the group as a whole. Figures 8.2, 8.3, and 8.4 are peer formats.

Figure 8.2   Peer Feedback for Cooperative Group Work (Integrates Domains)

Group member: _____

Feedback from: _____

How well did this group member contribute to work with this group? Circle the choices that match your thinking.

Action	Excellent		Acceptable		Needs to Improve
Started task work	Promptly		Needed a little push		Reluctant to start
Stayed on task	Almost always		OK but could improve		Seldom
Tried to answer own questions	Almost always		OK but could improve		Seldom
Helped keep noise level down	Almost always		OK but could improve		Seldom
Met deadlines	Almost always		OK but could improve		Seldom
Shared the workload	Almost always		OK but could improve		Seldom
Encouraged others	Almost always		OK but could improve		Seldom
Contributed ideas	Almost always		OK but could improve		Seldom
Cleaned up work area	Almost always		OK but could improve		Seldom
Was a team player	Almost always		OK but could improve		Seldom

Comments:

Note: Whether to have students provide assessments about other class members is a decision that the classroom teacher should make. Learning how to provide constructive feedback is an action that requires reflection and selection of words that can provide positive reinforcement. Where improvements are needed, the feedback task may be more of a challenge. Problems are possibilities for improvement, and the feedback from this form can be used to conference with students who need to improve in how they collaborate. This feedback form should be returned to the teacher, whose judgment can guide how to use the information.

Figure 8.3 Peer Feedback for Team Work (Integrates Domains)

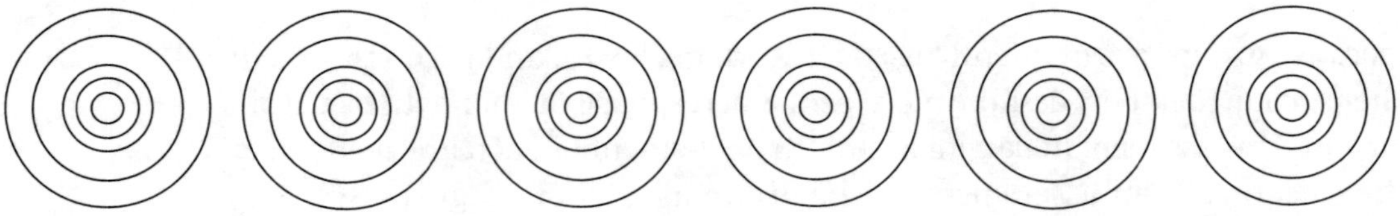

Team member: _____

Feedback from: _____

How well did this team member contribute to this team's work? Place a mark in the column that shows how well you think your team member was on target.

Action	Bull's-Eye	Hit the Target	Missed the Target
Started task work			
Stayed on task			
Tried to answer own questions			
Helped keep noise level down			
Met deadlines			
Shared the workload			
Encouraged others			
Contributed ideas			
Cleaned up work area			
A team player			

Something this team member did well was:

Something this team member could improve upon was:

Figure 8.4   Peer Feedback for Team Work (Integrates Domains)

Team member: _____

Feedback from: _____

How well did this team member contribute to this team's work? For each target action, mark an X where your team member was on the target.

Action	Action
Started task work 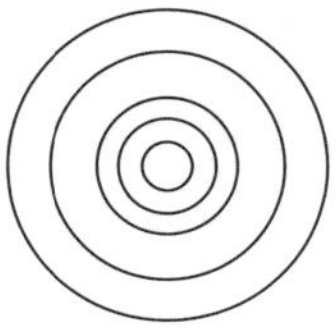	Shared the workload 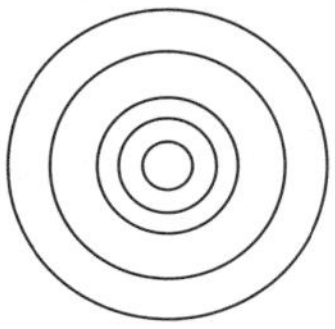
Stayed on task 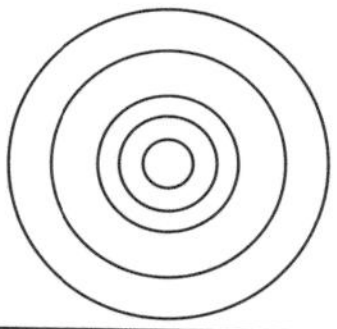	Encouraged others 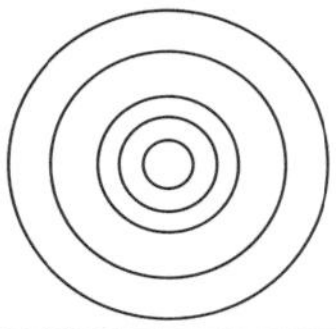
Tried to answer own questions 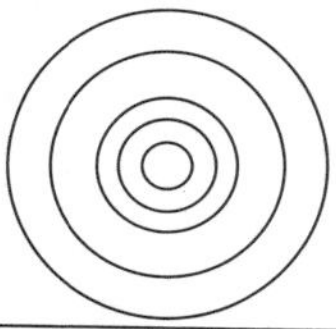	Contributed ideas 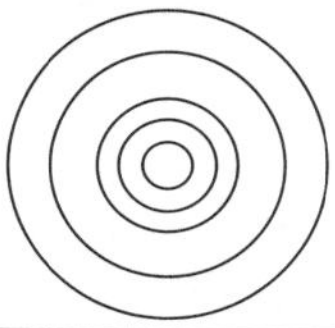
Helped keep noise level down 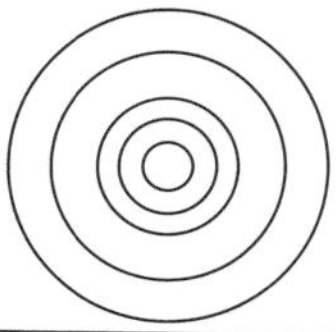	Cleaned up work area 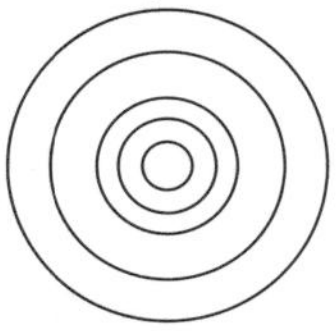
Met deadlines 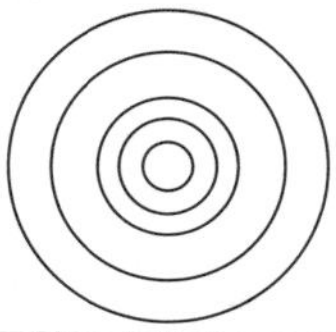	Overall, a team player 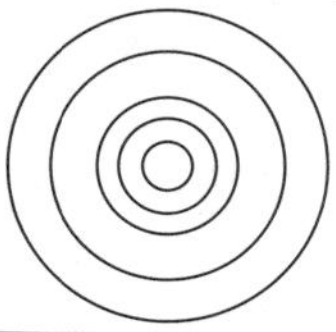

Something this team member did well was:

Something this team member could improve upon was:

CLASS ASSESSMENT OF GROUP PROJECT PRESENTATIONS

If students are completing presentations based on group work, both the teacher and the members of the class can assess the presentation. The presentation criteria should be shared with students in advance, and details such as the number and kinds of references required to support a project and presentation need to be communicated to students in the project criteria. The presentation criteria could also be made specific to the topic or concepts being studied. In the samples Figures 8.5 and 8.6, the numbers "4" and "2" are included in the event that the evaluation of the particular criterion falls in that borderline region.

Figure 8.5   Class Assessment of Group Project Presentation (Integrates Domains)

Project topic: _____

Group names: _____

Presentation Criteria

Criteria	Excellent—5	4	Acceptable—3	2	Needs Improvement—1
Understanding of the topic	Group knew what they were talking about		For the most part, group knew what they were talking about; a few errors present		Group hesitant at times; numerous errors were present
Questions about the topic	Group could answer all questions		Group could answer most questions		Group unable to answer most questions
Relatedness to unit of study	Project strongly supported unit of study		Project related to unit but lacked some connections		Project seemed vague and unrelated
Presentation skills	High interest; good eye contact; good delivery skills; very well organized		Moderate interest; some eye contact; delivery skills need some polish; good organization		Low interest; little eye contact; delivery skills need to be practiced; work needed in organization
Visual or media supports	Visuals or media supports enhanced the presentation		Some visuals or media supports included; need for more or better quality		Very limited or no supports provided
Participation of group members	All members had roles and participated		Most members had roles and participated		Very limited or no participation by some members
Research references	References; both primary and secondary sources used; exceeded project criteria		References; limited variety in types of sources; met minimum number		References very limited in types; did not meet project criteria

Score: _____/35 possible

What was a strength of the group presentation?

What is one suggestion you would make to improve the presentation?

Figure 8.6   Student Self-Assessment of Role in Group Project (Integrates Domains)

Project topic: _____

Name: _____

Self-Assessment Guide

Criteria	Excellent—5	4	Acceptable—3	2	Needs Improvement—1
What I learned	I learned many things about the topic.		I learned some things about the topic.		I didn't learn much about the topic.
My understanding of the concepts	I understand the concepts well enough to explain them to others.		I understand the concepts fairly well but still have some questions.		I had a great amount of difficulty understanding the concepts.
Identifying and finding resources and references	I was able to identify and find resources and references with little or no difficulty.		I had some difficulty finding these.		I was unable to find these.
My contributions	I often helped others with their tasks.		I occasionally helped others with their tasks.		I only had time to work on my own selected tasks.
Social skills	My group worked together very well.		My group had a few problems, but usually, we worked together quite well.		My group did not work together very well.
Doing my tasks	I finished all of my tasks on time.		I finished most of my tasks on time.		Some of my task work remained unfinished.

Score: _____/30 possible

My greatest contribution to the group was:

What I learned from working with this group on this project was:

The next time I work with a group, I will try to improve my:

9

Assessment Examples for Grades 9 Through 12

In this chapter, the suggestions and examples of assessments address the following:

- Science processes
- Understanding selected electricity concepts
- Understanding temperature and heat
- Understanding environmental sciences
- Food chains and food webs
- The student's view of the way science happens
- Understanding the nature of science
- Other question sources

USING THE STUDENT LABORATORY ENVIRONMENT INVENTORY

Laboratory teaching is an integral part of the science classroom. Although much is known about hands-on learning and many studies have corroborated that learning occurs best when one is *doing*, how can the effects of laboratory instruction on student learning and attitudes be measured? As mentioned in Chapter 5, Fraser, Giddings, and McRobbie (1992) designed the Student Laboratory Environment Inventory (SLEI) for situations in which a separate laboratory class exists. As noted in the earlier chapter, the SLEI integrates domains, consists of 35 items that are designed to measure five different dimensions, and uses the response alternatives of *almost never*, *seldom*, *sometimes*, *often*, and *very often*. Two SLEI forms can be used: The "Actual Form" asks students to note what actually takes place in the laboratory classroom, whereas the "Preferred Form" asks students to respond to what they would prefer to have take place in the laboratory class.

LOOKING AT SCIENCE PROCESSES

Students may use one or several process skills during an instructional period, and the teacher could use a predetermined observation checklist to document the use of process skills addressed in a particular class activity. Examples of items that could be used to assess some of the processes of science are provided in Figure 9.1. Students should be informed about the different performance levels and about what is considered an indicator of being able to use each process skill.

ASSESSING SPECIFIC AREAS OF UNDERSTANDING

Teaching for understanding can be a challenge—especially in science, where our intuitive descriptions of how things work are often not on target. As identified in *How Students Learn: Science in the Classroom* (National Research Council [NRC], 2005), students have preconceptions of how the world works. Teaching for understanding can be supported when the instructor knows how students think about concepts. Both knowing where students are (through preassessing) and knowing what the desired student outcomes are are essential. Preassessments should provide insights as to the kinds of learning experiences that help foster and develop understanding. The learning experiences that support the development of understanding very often are those experiences where students are engaged in doing science. Just doing something is not usually enough, but the doing must be supported with discussions and questioning.

A few examples of assessment instruments related to students' understandings of electricity, temperature and heat, environmental science, and food chains and food webs are set out in Figures 9.2–9.6.

Figure 9.1   Observation Checklist

Activity description: _____

Student: _____

Rating Scale

PP	LP	NMI
Proficient performance	Limited proficiency: Can use but needs more practice	Cannot do or use: Needs more instruction and practice

Circle the processes that apply to this activity:

Observing	Classifying	Communicating	Measuring
Predicting	Inferring	Variables—identify and control	Hypotheses—formulate and test
Interpreting data	Defining operationally	Experimenting	Constructing models

Other:

	PP	LP	NMI
1. Appropriate process skills were used while conducting the activity. Notes:	PP	LP	NMI
2. Appropriate process skills were used to complete and finish the activity. Notes:	PP	LP	NMI
3. Work was completed in the science notebook, journal, or log. Notes:	PP	LP	NMI
4. More practice is needed with these skills:			

Recommendations:

Figure 9.2 Environmental Science Concepts (Concept Domain)

Select the best answer.

1. A community is made up of clover, rabbits, and foxes. If an extended drought occurred in the area, the most immediate effect would likely be

   A. a decrease in the number of foxes.
   B. a decrease in the number of rabbits.
   C. a decrease in the amount of clover.*
   D. an increase in the number of rabbits.

2. The number of different species of birds, insects, and mammals occupying a given ecosystem is primarily dependent on which of the following?

   A. The amount of competition among species
   B. The diversity of vegetation in the system*
   C. The absence of large predators
   D. The time when vegetation is the thickest

3. Identify the type of area that would have these plants and animals: caribou, lemmings, lichens, moss, short grass, and wolves.

   A. The North American tundra*
   B. The eastern deciduous forest
   C. The southwestern desert
   D. A tropical rain forest

4. If the producers in an ecosystem were suddenly unable to use radiant energy from the sun, which of the following would immediately happen?

   A. Respiration in producers would suddenly cease.
   B. Photosynthetic activity would stop.*
   C. The system would increase its biomass.
   D. The decomposers in the system would cease to function.

5. What statement best describes the relationship between the number of producers in an ecosystem and the number of primary consumers in the same system?

   A. An increase in the number of producers is usually accompanied by an increase in primary consumers.*
   B. An increase in primary consumers is usually accompanied by an increase in producers.
   C. The number of producers in an ecosystem and the number of primary consumers in the same system are not related.
   D. The number of producers depends on the number of primary consumers.

Source: Adapted from Fleetwood, G. R. (1972). *Environmental Science Test.* Raleigh, NC: North Carolina Department of Public Instruction.

Figure 9.3   Preassessment of Understanding of Temperature and Heat (Concept Domain)

Check whether each statement is true or false. If you are not certain whether the statement is true or false, check "I'm not sure."

1. Temperature is the quantity of heat that is absorbed by an object.

   True _____   False _____   I'm not sure _____

2. When heat is absorbed by an object, its temperature always increases.

   True _____   False _____   I'm not sure _____

3. When heat is given off by an object, its temperature always decreases.

   True _____   False _____   I'm not sure _____

4. Temperature is a physical property of a substance. Some substances are cold and some are warmer.

   True _____   False _____   I'm not sure _____

5. Objects left inside a container for a long time will all reach the same temperature.

   True _____   False _____   I'm not sure _____

6. If the temperature of an object that is heated remains the same, then there is probably a change in the state of the object.

   True _____   False _____   I'm not sure _____

7. A larger piece of ice has a lower temperature than a smaller one; therefore, it melts more slowly.

   True _____   False _____   I'm not sure _____

8. Temperature is a kind of energy. Objects with higher temperatures contain more heat, and objects with lower temperatures contain less heat.

   True _____   False _____   I'm not sure _____

Please write down other things that you know or do not know about temperature or heat.

Source: Adapted from Guo, C. J. (1993). *Alternative Frameworks of Motion, Force, Heat, and Temperature: A Summary of Studies for Students in Taiwan.* Paper presented at the annual meeting of the National Association for Research in Science Teaching, Atlanta, Georgia.

Figure 9.4   Applying Concepts of Heat: Preassessment for Concepts of Heat Transfer (Application Domain)

Ask each student to draw a picture of an ice cube in a glass of water. All of the students should then draw arrows to show the directions of heat flow. Students should also provide explanations to support their drawings.

A recommendation: Support this activity by having the actual materials and thermometers for students to take measurements while they are thinking and drawing.

Next, ask each student to draw a picture of a lighted candle. All of the students should then draw arrows to show the directions of heat flow. Students should also provide explanations to support their drawings.

A recommendation: Support this activity by having candles so that students can make observations while they are thinking and drawing.

After students have completed their drawing and writing, they could compare and contrast their answers, and the instructor could mentally note misunderstandings on which to focus instruction. A postassessment could ask students to respond to these same kinds of situations. In checking for understanding, the students could be asked to make drawings and add descriptions. The students' preassessment drawings could be returned and students could make comparisons and indicate how their thinking has changed.

Figure 9.5   Energy Efficiency Assessment (Application Domain)

A basic floor plan for a house is set out, and although you plan to have air conditioning, you want to keep your utility bills as low as possible. Show where you would locate trees around the yard in order to help conserve energy. Assume that the house is in an area where you can have snow in the winter and temperatures as high as in the 90s Fahrenheit in summer. Explain why you would place the trees where you did.

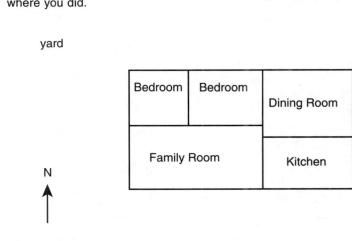

Figure 9.6   Food Chains and Food Webs Questions (Application Domain)

When one organism becomes food for another organism and energy is transferred via eating and being eaten, the flow of energy is called a *food chain*. This chain began when the producers absorbed the sun's energy to produce their own food. Then consumers ate the producers to obtain their energy, and those consumers may also have been eaten by other consumers. The connectivity of feeding relationships can be depicted as a food web.

A Food Web

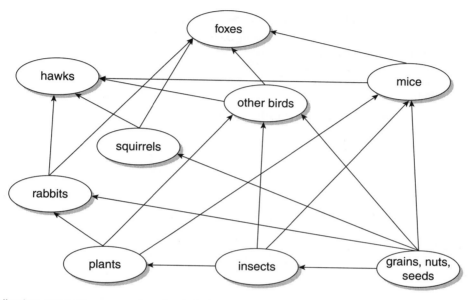

The following example represents a food chain:

   Plants → Insects → Mouse → Hawk

Eating and being eaten becomes a much more complex process in the ecosystem, and food chains begin to interconnect to form food webs, as in the example shown in the food web above.

Using the information presented about food chains and food webs, answer the questions that follow.

1. When a large group of cats was introduced to the food web shown, almost all of the mice were eaten in just a few weeks. Which organisms would be most likely to have the largest increase in population as a result?
   A. The insects*
   B. The foxes
   C. The rabbits
   D. The hawks
   E. The squirrels

2. The decomposers, organisms that obtain their food from both wastes and dead organisms, are not shown in the food web. Decomposers are important in the food web because they do which of the following?
   A. Recycle nutrients*
   B. Inhibit diseases
   C. Control population growth
   D. Compete with producers
   E. Compete with consumers

(Continued)

Figure 9.6 (Continued)

3. Actually, the food web shown is a hypothetical example. In reality, it could be expected that food webs would be

    A. very similar to the illustration.
    B. more complex than the illustration.*
    C. simpler than the illustration.
    D. uncommon in nature.
    E. nonexistent in nature.

4. Based on the information provided, which of the following statements is not correct?

    A. Each level of the food chain depends on the level below.
    B. Energy travels in both directions in a food chain.*
    C. The food web shows many interrelated food chains.
    D. Consumers often compete for the same food source.
    E. Most consumers eat a variety of foods.

5. Which of the following would be a producer in the ocean?

    A. Shrimp
    B. Mussels
    C. Algae*
    D. Salt
    E. Whales

6. In the food chain, consumers are ranked according to

    A. their size.
    B. their weight.
    C. where they live.
    D. what they eat.*
    E. how much they eat.

ASSESSING STUDENTS' VIEWS OF THE SCIENTIFIC PROCESS

Figures 9.7 and 9.8 are examples of student self-assessment instruments regarding the Nature of Science Domain.

Other Question Sources

The Field-Tested Learning Assessment Guide (FLAG) (http://www.flaguide.org/intro/intro.php) has various instruments available, including the Views of Nature of Science Questionnaire (VNOS) that has three open-ended versions. The most frequently used versions are the VNOS–B (seven items) and the VNOS–C (ten items), and these instruments can be accessed online. Each instrument is designed to elucidate students' views about several aspects of the nature of science (NOS). Background information about the development of VNOS–B and VNOS–C in order to assess students' conceptions of the nature of science was described by Lederman, Abd-El-Khalick, Bell, & Schwartz (2002). The VNOS instruments are intended to ascertain how students and others perceive the explanatory nature of

Figure 9.7   What You Think About the Nature of Science (Nature of Science Domain)

Student name: _____

For each question, circle the letter that best indicates your viewpoint.

	Strongly Agree	Agree	Disagree	Strongly Disagree
1. Science means questioning, explaining, and testing.	a	b	c	d
2. Science means studying the concepts developed and known by scientists.	a	b	c	d
3. Science means working with various objects and materials in classrooms and laboratories.	a	b	c	d
4. Science deals with activities that affect living: in homes, schools, communities, and nations.	a	b	c	d
5. Science is a human activity that involves acting on questions about the universe.	a	b	c	d
6. Science is a body of knowledge, developed over years, about the universe.	a	b	c	d
7. Science can be defined by what scientists know.	a	b	c	d
8. Science is an attempt to know more about the world around us.	a	b	c	d
9. Science is a way of viewing the universe and how it works.	a	b	c	d
10. Science is based on curiosity about objects and events in the universe.	a	b	c	d
11. Science is based on attempts to answer the questions about the objects and events in the universe.	a	b	c	d
12. Science must include tests in nature that illustrate the validity of the personal explanations offered.	a	b	c	d
13. Science is a self-correcting human endeavor.	a	b	c	d
14. Any theory or concept of science can be challenged.	a	b	c	d

Figure 9.8 Your View of the Way Science Happens (Nature of Science Domain)

Student name: _____

For each question, mark the letter that indicates your view of the best response.

1. The main reason that scientists study plants is

   A. to teach farmers how to grow more food.
   B. to learn how to make better medicine from plants.
   C. to be able to explain how plants grow.*
   D. to be able to tell what the best soil is for plants.

2. Over 200 years ago, a scientist named Sir Isaac Newton explained the motion of planets with what is called *gravity*. Scientists decided that Newton's explanation was right. Not long ago, a scientist named Albert Einstein said that his idea (relativity) could explain everything that Newton's idea (gravity) could explain, plus more. Today, scientists accept Einstein's idea (relativity). What do they say about Newton's idea?

   A. Newton's idea was wrong, but Newton didn't study as many things as Einstein did.
   B. Newton's idea was right, but Newton's idea can't explain as much as Einstein's idea can.*
   C. Newton's idea will work only if you are on a planet somewhere else in space.
   D. Newton's idea is better than Einstein's because it has been used longer.

3. A scientist in Australia says that she has seen signs of plant life on the planet Venus. American scientists will believe this scientist

   A. if other scientists also see these signs of plant life.*
   B. if she can tell them what type of plants these probably are.
   C. if the Australian government will stand behind the scientist's work.
   D. if other scientists agree that there is oxygen on Venus.

4. If we ask an astronomer to explain why some stars look brighter than others, her or his explanation would

   A. tell us why it is necessary for stars to differ in brightness.
   B. use scientific laws.*
   C. be made with mathematical equations and formulas.
   D. use observations and data from astronomy.

5. Science is an attempt to

   A. make sure that what has been discovered about the world is really true.
   B. give us laws and theories that help explain nature.*
   C. discover, collect, and group facts about nature.
   D. find ways to make people's lives better.

6. Which of these statements best describes scientific knowledge?

   A. Scientific knowledge is a well-organized collection of facts.
   B. Today's scientific knowledge builds on data and ideas from the past.*
   C. Today's scientific knowledge has been produced by today's scientists.
   D. Scientific knowledge contains only statements that are 100% true.

7. Are biology, chemistry, and physics related to each other?

   A. They are not related because they are built on different basic ideas.
   B. They are related because some of the ideas in each relate to the ideas in the others.*
   C. They are related because mathematics ties them together.
   D. They are not related because each one studies something very different from what is studied by the others.

8. Today, scientists are doing experiments to see if one of Einstein's theories is right when it predicts that light rays will bend as they pass near a very large object, such as a star. This is an example of

   A. how theories suggest experiments to try.*
   B. how it is important to have an exact measurement of how fast light travels.
   C. how experiments must be done to prove that a theory is really true.
   D. how theories are still doubted long after they are proven true.

9. Designing a television set is a problem for
   A. science because to design a set, you need brain power.
   B. science because to design a set, you need to do experiments.
   C. technology because it leads to a useful device.*
   D. technology because it involves working with electricity.

10. Betty is planning an experiment to find out whether potassium is important in the growth of a certain plant. Her teacher suggests growing one group of these plants in soil with nitrogen and phosphorus but no potassium. The teacher will probably also suggest that Betty grow another group in soil with

    A. potassium only.
    B. nitrogen, phosphorous, and potassium.*
    C. nitrogen and potassium but with no phosphorus.
    D. nitrogen and phosphorus but with no potassium.

11. All of the following statements deal with how science and technology are connected. Which statement best describes the connection?

    A. Technology uses scientific knowledge to help solve practical problems.*
    B. Science depends on technology for ideas and for help in planning experiments.
    C. The laws used in science come from technology.
    D. Technology is the part of science that solves mechanical problems.

12. John was asked to describe a scientific law. Which of the following descriptions should he give?

    A. A scientific law is an exact report of what scientists observe.
    B. A scientific law says how one type of event in nature is related to another event.*
    C. A scientific law is an explanation of an event in nature, and it uses things that can't be seen.
    D. A scientific law is a rule that nature makes, and it cannot be broken.

13. A certain law cannot explain some of the facts that it should explain. What may scientists do with this law?

    A. Scientists may change the unexplained facts so that the law can explain them.
    B. Scientists may change the law so that more of the facts can be explained.*
    C. Scientists should throw out the law and make a new one immediately.
    D. Scientists should show that the law is wrong in the case of all facts.

14. Robert Hooke did many experiments with springs and found that most springs stretch by twice as much when the amount of weight hung on the spring is made twice as great. Springs stretch by three times when the amount of weight is made three times as great, stretch four times as much with four times as much weight, and so on. This led Hooke to make a general rule for how much springs stretch when different weights are hung from them. This story is an example of how some scientists

    A. have come up with a scientific theory.
    B. have tested a scientific hypothesis.
    C. have come up with a scientific law.*
    D. have come up with ideas about real things from ideas about imaginary things.

15. When a new theory is suggested, scientists will probably decide that it is a good theory
    A. if they think the theory is true.
    B. if the theory can be put into a mathematical equation.
    C. if the theory fits with what they observe and agrees with their other ideas.
    D. if the theory fits with observations.*

(Continued)

Figure 9.8 (Continued)

16. When scientists decide that a new theory is an acceptable theory, we can say that

    A. science's ideas and explanations of nature have increased.*
    B. one more of the laws of nature is now known.
    C. we are closer to the end of the search for scientific knowledge.
    D. science has discovered new experimental evidence.

17. Many people say that scientists behave in special ways. For example, they observe carefully, they do not jump to conclusions, and they are very exact. If we wanted to see scientists behave this way, it would be best to watch them when

    A. they are doing experiments.*
    B. they are doing work outside of science.
    C. they are doing almost anything.
    D. they are with their family and friends.

18. In discussing the problem of nuclear weapons, a famous scientist said that we must keep on doing experiments with nuclear weapons. The scientist is probably

    A. right because his or her scientific attitude makes the answers to problems more correct.
    B. wrong because scientists seem to be trying to destroy the world.
    C. right because scientific results are right more often than other kinds of results.
    D. no more right or wrong than any other intelligent person, unless he or she studied the problem.*

19. A book about science says, "Scientists do experiments to ask nature questions." This means that experiments are used in science

    A. to prove that nature follows rules.
    B. to learn by trying different solutions to a problem until one works.
    C. to see if predictions made from scientists' ideas are right.*
    D. to try to find out about the origin of humans.

20. Lisa has a good imagination, but she may never become a scientist because

    A. she would not want to give up being able to think anything that she wants to think.
    B. people with a good imagination usually become artists and writers.
    C. she might like some other field better than science.*
    D. science has too many facts.

21. When scientists carefully measure any quantity many times, they expect that

    A. all of the measurements will be exactly the same.
    B. only two of the measurements will be exactly the same.
    C. all but one of the measurements will be exactly the same.
    D. most of the measurements will be close but not exactly the same.*

22. Which of the following is the description of a scientific theory?

    A. A scientific theory uses arithmetic.
    B. A scientific theory describes a scientist.
    C. A scientific theory describes an experiment.
    D. A scientific theory helps explain why some things act the way they do.*

Source: Adapted from Cooley, W. W., and Klopfer, L. E. (1961). *Test on Understanding Science.* Princeton, NJ: Educational Testing Service.

science and how scientific understandings develop. The open-ended questions elicit thinking centered on the following ideas.

- Science is based, at least partially, on observations of the natural world. Observation drives science.
- Scientific knowledge and investigation are theory laden.
- Scientific knowledge is tentative and subject to change.
- Science is evidence-based.
- Scientific knowledge involves imagination and creativity.
- Science is a human endeavor and is influenced by culture.
- No single scientific method exists. Scientists use systematic ways of investigation.
- Theories in science are overarching and explanatory.

Released items from the National Assessment of Educational Progress (NAEP) can be accessed online at http://nces.ed.gov/NATIONSREPORTCARD. With the "Sample Questions" tab, you can navigate to the page with "Questions Tool," and with this tool you can select from released science items for 2000 and 2005. The next administration of the NAEP science test is in Spring 2009, and this test will also include performance tasks. Sample items are available for Grades 4, 8, and 12. Based on student response data, items are rated as easy, medium, or hard, and the items formats are multiple-choice or constructed response. Scoring guidelines are available, and sample answers are posted with raters' comments. The sample items can be used to create assessments for formative and summative purposes. Single questions might be selected or a "testlet" could be created for checking student understanding of concepts in a science module or unit.

Glossary of Assessment-Related Terminology

Application Domain

The key question is, To what extent can students transfer and effectively use what they have learned to a new situation, especially in daily life (Grönlund, 1988)? Science-Technology-Society (STS) stresses the association of scientific knowledge with students' social and living experiences by using current social issues to help students see their connections. Starting from concerns in the real world may be a way to diminish the learning gap between the two worlds of school-science experiences and personal technology experiences (Yager & McCormack, 1989).

Assessment

The key questions are, What do students know? How well do teachers teach? Do our schools work? (Kulm & Malcom, 1991).

Alternative assessment applies to assessments that differ from the multiple-choice, timed, one-shot approaches that characterize most standardized and many classroom assessments (Marzano, Pickering, & McTighe, 1993). The reason to promote the use of alternative assessment is to avoid the drawback of ignoring performances other than outcomes of standardized tests.

Authentic assessment conveys the idea that assessment should engage students in applying knowledge and skills in the way they are used in the real world. The key point is "test students in context" rather than by using the standardized tests. It also reflects good instructional practice, so that teaching to the test is desirable (Wiggins, 1989a, 1989b).

Embedded assessment is assessment that is incorporated into instruction. When learning experiences are designed, assessment opportunities are also planned and designed.

Formative assessment is assessment that is used with the intent to improve instruction and students' learning (Brookhart & Nitko, 2008).

Holistic assessment encompasses the totality of student performance in the classroom, and like performance assessment and beyond-test outcomes, many components of student learning are important to assess. An excellent way for a teacher to evaluate student work in a classroom or laboratory is to use an observational rubric to record observations and comments about student performance. Observations help the teacher note not only when students are successful, but also help identify students who may have difficulty with equipment usage, those who contribute ideas, and those who hesitate to participate (Raizen & Kaser, 1989).

Performance assessment refers to the variety of tasks and situations in which students are given opportunities to demonstrate their understanding and to thoughtfully apply knowledge, skills, and habits of mind in a variety of contexts (Marzano et al., 1993).

Summative assessment is assessment that is used to make judgments about students' learning after an instructional process is completed (Brookhart & Nitko, 2008).

Attitude Domain

This is a state of mind or a feeling. The term is very broadly used in discussions about science education, but often with various meanings. It is possible to distinguish two broad categories: attitude toward science (i.e., interest in science, attitude toward scientists, or attitudes toward social responsibility in science) and scientific attitude (i.e., open-mindedness, honesty, or skepticism) (Gardner, 1975). Whereas *Science for All Americans* (American Association for the Advancement of Science [AAAS], 1990) postulates scientific literacy for all Americans, STS curriculum tries to promote students' positive attitudes toward science, which includes an I-can-do-it attitude and decision making about social and environmental issues (Yager & McCormack, 1989).

Behaviorist Teaching Approach

In the behaviorist approach to teaching, Mayer (1987) noted that what is of interest to the teacher is the relationship between instructional manipulations and the outcome performance. Behaviorist approaches tend to focus more on overt behaviors, whereas cognitive approaches focus on both overt and covert behaviors.

Cognitive Teaching Approach

The cognitive approach attempts to understand (a) how instructional manipulations affect internal cognitive processes, such as paying attention; (b) how these processes result in the acquisition of new knowledge; and (c) how new knowledge influences performance, such as on tests. The goal is to explain the relationship between stimulus and response by describing the intervening cognitive processes and structures (Mayer, 1987).

Concept Domain

Conceptual systems are primarily structured via *kind* or *is-a* hierarchies (e.g., Tweety is a canary, which is a kind of bird, which is a kind of animal, which is a kind of organism) and *part-whole* hierarchies (e.g., a toe is part of a foot, which is part of a leg, which is part of a body) (Thagard, 1992). Helping students to construct the natural world into their own categorization in terms of scientific knowledge is one of the basic functions of science education. This domain includes facts, laws (principles), theories, and the internalized knowledge of students (Yager & McCormack, 1989).

Concept Mapping

Concept mapping is a technique for externalizing concepts and propositions. Propositions are two or more concept labels linked by words in a semantic unit. Concept maps are an explicit, overt representation of the concepts and propositions a person holds. They allow teachers and learners to exchange views on why a particular propositional linkage is good or valid or to recognize missing linkages between concepts (Novak & Gowin, 1984). Concept mapping is always used with interviews to sort out the underlying meaning of each proposition.

Constructivism

The learner constructs and frames meaning in terms of his or her existent experiences in a social context. Knowledge is not passively received but is actively built upon past experiences. Cognition is adaptive and helps the learner organize the experiential world, not the discovery of ontological reality. The learner does not find truth but instead constructs viable explanations based on personal experiences (Wheatley, 1991).

Cooperative Learning

It could be argued that students should learn through discussion in social contexts and that personal learning cannot be independent from external social activities. Working in small groups can provide students with opportunities to challenge one another's ideas, and this can lead them to realize the need to reorganize and reconceptualize their thinking. Johnson and Johnson (1990) provided evidence that when students learned more social skills, their performance on standardized math computation scales became better.

Creativity Domain

Torrance (as cited in Penick, 1996) described creativity as a process of becoming sensitive to problems, deficiencies, gaps in knowledge, missing elements, and disharmonies. Creativity involves the identification of the problematic; the search for solutions, the generation of guesses, or the formulation of hypotheses about the deficiencies; testing and retesting the hypotheses and the potential modification and retesting; and last, the communication of the results. STS teaching strategies stress the role of guessing as a basic element of creativity.

When students view natural phenomena, to stimulate their creativity and interpretation of results, they should be encouraged to present wild guesses to generate diverse perspectives (Yager & McCormack, 1989).

Criterion-Referenced

Criterion-referenced is used to compare a student's performance against a performance domain for which criteria have been established. Evaluation of performance is determined based upon how well a student meets criteria. Performance levels are often used to communicate results.

Evaluation

Evaluation is a systematic process of collecting, analyzing, and interpreting information for judging decision alternatives. It not only uses quantitative measurements and, in some cases, qualitative information, but also involves value judgments. In the classroom, evaluation is the process of interpreting students' performances for both formative and summative purposes.

Formal Assessments

Formal assessments can vary in format and are usually scheduled and well planned. Formal assessments could range from scheduled examinations in classes throughout a grading period to a defense of a thesis. Formal assessments can include performances, portfolios, interviews, multimedia work, and other examples. Standardized tests are formal assessments. What makes the assessment formal also rests in the types of decisions to be made based on the assessment. Just like with formal attire, a continuum exists with the level of formality.

Formal Tests

Formal tests tend to be those placed at end points. Time parameters and standardized instructions are in place. State assessments and other standardized tests fall in this category.

Informal Assessments

Informal assessments are those assessments that, while often planned, tend to be used for getting a read on where students are in their understandings. Examples of informal assessments include questioning, preassessments, observations, and listening to student conversations while they are working. The assessment information may exist, for example, as mental notes, observations, preassessment data, or checklists. Informal assessments are often assessments that are done as a teacher is walking around a classroom. Informal assessments inform instruction, and they tend to be formative in nature.

Informal Test

An *informal test* is one constructed by a teacher for use on one or a limited number of occasions where comparability or results across groups is not essential.

Law

In science, *laws* are often expressed as equations relating measurable parameters. Laws indicate the mathematical relations that hold true among quantities of certain empirical parameters. Laws provide us with knowledge of facts but not with an explanation as to why the facts are as they are (Dilworth, 1981).

Measurement

Measurement is the process of assigning numbers to attributes of individuals according to specified rules (it always refers to quantification).

Nature of Science Domain

STS stresses the nature of the development of scientific knowledge and the influence coming from the external social factors, which in turn discount the absolute objectivity of science. Abimbola (1983) identified these basic themes about the nature of science: Observations are theory laden, and the existing paradigms determine problem selection, instrumentation, and the inferential techniques and models used. The scientific community ultimately drives the choice of the scientific question, and formal logic is replaced by a reliance on the detailed study of the history of science. Continued research and the accompanying critique by the scientific community are at the core of science. Science progresses in two phases: normal science and revolutionary science. Normal science operates in the context of a shared paradigm and is responsible for generating scientific revolutions. The more important events in the history of science are those revolutions that change paradigms. Progress in science is, therefore, noncumulative because of paradigm shifts. Observational data do not remain the same from one scientific revolution to another because scientific paradigms are incommensurable. Scientists view events through the lens of a new paradigm.

As Lederman (1992) suggested, the students' conception of the nature of science was inadequate, and Carey, Evans, Honda, Jay, and Unger (1989) suggested that the teaching of the nature of science can improve learning. Other researchers (e.g., Matthews, 1991) have suggested that the use of historical case studies may be one way to build greater student understanding about the nature of science.

Norm-Referenced

Norm-referenced is used to compare a student's performance to that of a norm group that is a well-defined comparison group that sat for the same assessment under very specific or controlled conditions.

Process Domain

Practical work, including hands-on activities, scientific inquiries, or experiments, is always cited as the most powerful approach to helping students understand scientific knowledge. At the same time, Hodson (1992) argued that the skills-based approach of practical work was philosophically unsound, educationally worthless, and pedagogically dangerous. STS suggests a holistic view of assessment to promote valid process learning, especially emphasizing the role of creativity in the data analysis process (Yager & McCormack, 1989).

Reliability

Reliability is the consistency or the degree of consistency between two or more measures of the same thing. The key questions are (a) Are the performances of students consistent? and (b) To what degree? High reliability of a test does not necessarily indicate a good measurement unless there is also high validity.

Rubrics, or Scoring Guides

Rubrics, or scoring guides, are used to score nontraditional assessments. A typical rubric (a) contains anywhere from 3 to 10 levels of performance, (b) states all the major dimensions to be assessed, and (c) provides necessary information about idiosyncrasies in the question or equipment. Holistic, analytic, or component rubrics can be used.

Scientific Inquiry

Scientific inquiry is the process in which phenomena are investigated and the results of observations are interpreted. From a positivist viewpoint, students can reach scientific knowledge without help or external guidance. Yet more and more research confirms that students' prior knowledge may bias what they observe and create misconceptions (Matthews, 1988). According to Ausubel (1968), the most important single factor influencing learning is what the learner already knows. If this can be ascertained, then the instruction should reflect what the learner already knows.

Standardized Test

A *standardized test* is typically constructed for use in more than one setting. It is standardized in the sense that the administrative procedures, directions, apparatus, and scoring are fixed by the constructors so that the test may be administered and scored identically by different examiners in different settings to achieve comparable results across all examined.

Test

A *test* is a systematic procedure for observing behavior and describing it with the aid of numerical scales or developed categories.

Theory

The primary function of a *theory* is to provide the conception of a mechanism that can explain the empirical regularities behind appearances, as do laws. Yet theories simply offer potentially true descriptions of a reality. Thus, a theory can one day be abandoned as a result of measurement made by ever more sensitive instruments or the arrival of a superior alternative (Dilworth, 1981).

Validation

Validation is "the process by which a test user collects evidence to support the types of inferences that are to be drawn from test scores (Crocker & Algina, 1986, p. 217).

Validity

"The degree to which the test actually measures what it purports to measure" (Anastasi, 1982, p. 28) is called *validity*. The key questions are (a) Did we test what we wanted to test? and (b) To what degree? High validity usually indicates high reliability.

References

Abimbola, I. O. (1983). The relevance of the "new" philosophy of science for the science curriculum. *School Science and Mathematics, 83*(3), 182–190.

Aikenhead, G. S. (1973). The measurement of high school students' knowledge about science and scientists. *Science Education, 57*(4), 539–549.

Aikenhead, G. S. (1979). Science: A way of knowing. *The Science Teacher, 46*(6), 23–25.

Akindehein, F. (1988). Effect of an instructional package on pre-service science teachers' understanding of the nature of science and acquisition of science-related attitudes. *Science Education, 72*(1), 73–82.

American Association for the Advancement of Science. (1968). *Science: A process approach.* Washington, DC: Author.

American Association for the Advancement of Science. (1990). *Science for all Americans.* Washington, DC: Author.

American Association for the Advancement of Science. (1993). *Benchmarks for science literacy.* Washington, DC: Author.

American Association for the Advancement of Science. (2001). *Atlas of scientific literacy: Volume 1.* Washington, DC: Author.

American Association for the Advancement of Science. (2007). *Atlas of scientific literacy: Volume 2.* Washington, DC: Author.

Anastasi, A. (1982). *Psychological testing.* New York: Macmillan.

Angelo, T. A., & Cross, K. P. (1993). *Classroom assessment techniques: A handbook for college teachers.* San Francisco: Jossey-Bass.

Arter, J. A., & Chappuis, J. (2006). *Creating & recognizing quality rubrics.* Portland, OR: Educational Testing Service.

Arter, J. A., & McTighe, J. (2001). *Scoring rubrics in the classroom: Using performance criteria for assessing and improving student performance.* Thousand Oaks, CA: Corwin.

Ausubel, D. P. (1968). *Educational psychology: A cognitive view.* New York: Holt, Rinehart & Winston.

Barenholz, H., & Tamir, P. (1992). A comprehensive use of concept mapping in design instruction and assessment. *Research in Science and Technological Education, 10*(1), 37–52.

Barnum, C. R. (1996). How do students really view science and scientists? *Science and Children, 34*(1), 30–33.

Barnum, C. R. (1997). Students' view of scientists and science: Results from a national study. *Science and Children, 354*(1), 18–24.

Baron, J. B. (1991). Strategies for development of effective performance exercises. *Applied Measurement in Education, 4*(4), 305–318.

Barufaldi, J. P., Bethel, L. J., & Lamb, W. G. (1977). The effect of a science methods course on the philosophical view of science among elementary education majors. *Journal of Research in Science Teaching, 14*(4), 289–294.

Black, P., & Wiliam, D. (1998). Inside the black box: Raising standards through classroom assessment. *Phi Delta Kappan, 80*(2), 139–148.

Brookhart, S. M. (2008). *How to give effective feedback to your students.* Alexandria, VA: Association for Supervision and Curriculum Development.

Brookhart, S. M., & Nitko, A. J. (2008). *Assessment and grading in classrooms.* Upper Saddle River, NJ: Pearson Education.

Brooks, J. G., & Brooks, M. G. (1993). *In search of understanding: The case for constructivist classrooms.* Alexandria, VA: Association for Supervision and Curriculum Development.

Burry-Stock, J. A. (1993). *Expert science teaching evaluation model (ESTEEM) training manual.* Kalamazoo: Western Michigan University, The Evaluation Center.

Carey, S., Evans, R., Honda, M., Jay, E., & Unger, C. (1989). An experiment is when you try it and see if it works: A study of Grade 7 students' understanding of the construction of scientific knowledge. *International Journal of Science Education, 11*, 514–529.

Champagne, A. B., & Newell, S. T. (1992). Directions for research and development: Alternative methods of assessing scientific literacy. *Journal of Research in Science Teaching, 29*(8), 841–860.

Chesbro, R. (2006). Using interactive science notebooks for inquiry-based science. *Science Scope, 29*(7), 30–34.

Cooley, W. W., & Klopfer, L. E. (1961). *Test on understanding science.* Princeton, NJ: Educational Testing Service.

Crocker, L., & Algina, J. (1986). *Introduction to classical and modern test theory.* New York: CBS College Publishing.

Csikszentmihalyi, M. (1990). *Flow: The psychology of optimal life experience.* New York: HarperCollins.

Csikszentmihalyi, M. (1996). *Creativity: Flow and the psychology of discovery and invention.* New York: HarperCollins.

Dilworth, C. (1981). *Scientific progress.* Dordrecht, Holland: Dordrecht Publishers.

Douglas, R., Klentschy, M. P., & Worth, K. (Eds.). (2006). *Linking science & literacy in the K–8 classroom.* Arlington, VA: NSTA Press.

Edwards, B. (1979). *Drawing on the right side of the brain.* Los Angeles: J. P. Tarcher.

Enger, S. K. (1997). *The relationship between science learning opportunities and ninth grade students' performance on a set of open-ended science questions.* Unpublished doctoral dissertation, The University of Iowa, Iowa City.

Felker, E. (1974). *Building positive self-concepts.* Minneapolis, MN: Burgess.

Fleetwood, G. R. (1972). *Environmental science test.* Raleigh, NC: North Carolina Department of Public Instruction.

Fort, D. C., & Varney, H. L. (1989). How do students see scientists: Mostly male, mostly white, and mostly benevolent. *Science and Children, 26*(8), 8–13.

Fraser, B. J., Giddings, G. J., & McRobbie, C. J. (1992). Assessing the climate of science laboratory classes. *What Research Says to the Science and Mathematics Teacher, 8*, 1–8.

Gallagher, J. J. (1999). Improving science teaching and student achievement through embedded assessment. *Michigan Science Teachers Association, 144*(3). Retrieved December 21, 2008, from http://www.msta-mich.org/index.php/publications/journalArticle/20

Gardner, P. L. (1975). Attitudes to science: A review. *Studies in Science Education, 2*, 1–41.

Gay, L. R., & Airaisian, P. (2000). *Educational research: Competencies for analysis and application.* Upper Saddle River, NJ: Prentice Hall.

General Educational Development. (1990). *The 1988 tests of general educational development: A preview.* Washington, DC: American Council on Education.

Giddings, G. (1993). *Student instruction and motivation survey.* Perth, West Australia: Curtin University.

Gilbert, J., & Kotelman, M. (2005). Five good reasons to use science notebooks. *Science and Children, 43*(3), 28–32.

Grönlund, N. E. (1988). *How to construct achievement tests.* Englewood Cliffs, NJ: Prentice Hall.

Grönlund, N. E., & Linn, R. L. (1990). *Measurement and evaluation in teaching.* New York: Macmillan.

Guo, C. J. (1993). *Alternative frameworks of motion, force, heat, and temperature: A summary of studies for students in Taiwan.* Paper presented at the annual meeting of the National Association for Research in Science Teaching, Atlanta, GA.

Herman, J. L., Osmundson, E., Ayala, C., Schneider, S., & Timms, M. (2006, December). *The nature and impact of teachers' formative assessment practices* (CSE Report 703). Los Angeles: UCLA, National Center for Research on Evaluation, Standards, and Student Testing (CRESST).

Hodson, D. (1992). Assessment of practical work: Some considerations in philosophy of science. *Science & Education, 1*(2), 115–144.

Hodson, D., & Reid, D. J. (1988). Science for all: Motives, meanings and implications. *The School Science Review, 69*(249), 653–661.

Hopkins, D. (1993). *A teacher's guide to classroom research.* Buckingham, UK: Open University Press.

Horvath, F. G. (1991). *Assessment in Alberta: Dimensions of authenticity.* Paper presented at the annual meeting of the National Association of Test Directors and the National Council on Measurement in Education, Chicago, IL.

Jellen, H. G., & Urban, K. K. (1986). The TCT-DP (test for creative thinking-drawing production): An instrument that can be applied to most age and ability groups. *The Creative Child and Adult Quarterly, 11*(3), 138–144.

Johnson, D. W., & Johnson, R. T. (1983). Interdependence and interpersonal attraction among heterogeneous and homogeneous individuals: A theoretical formulation and a meta-analysis of the research. *Review of Educational Research, 53*(1), 5–54.

Johnson, D. W., & Johnson, R. T. (1990). Social skills for successful group work. *Educational Leadership, 47*(4), 29–33.

Jones, R. W. (1994). *Performance and alternative assessment techniques: Meeting the challenges of alternative evaluation strategies.* Paper presented at the Second International Conference on Educational Evaluation and Assessment, Pretoria, Republic of South Africa.

Klentschy, M. P. (2008). *Using science notebooks in elementary classrooms.* Arlington, VA: NSTA Press.

Knorr-Cetina, K. D. (1981). *The manufacture of knowledge: An essay on the constructivist and contextual nature of science.* New York: Pergamon.

Krajcik, J., & Czerniak, C. (2007). *Teaching science in elementary and middle school: A project-based approach.* New York: Lawrence Erlbaum.

Kuhn, T. S. (1962). *The structure of scientific revolutions.* Chicago: University of Chicago Press.

Kulm, G., & Malcom, S. M. (1991). *Science assessment in the service of reform.* Washington, DC: American Association for the Advancement of Science.

Lederman, N. G. (1992). Students' and teachers' conceptions of the nature of science: A review of the research. *Journal of Research in Science Teaching, 29,* 331–359.

Lederman, N. G., Abd-El-Khalick, F., Bell, R. L., & Schwartz, R. (2002). Views of nature of science questionnaire: Toward valid and meaningful assessment of learner's conceptions of nature of science. *Journal of Research in Science Teaching, 39*(6), 497–521.

Marzano, R. J., Pickering, D., & McTighe, J. (1993). *Assessing student outcomes: Performance assessment using the dimensions of learning model.* Alexandria, VA: Association for Supervision and Curriculum Development.

Matthews, M. R. (1988). A role for history and philosophy in science teaching. *Educational Philosophy and Theory, 20*(2), 67–75.

Matthews, M. R. (1991). *History, philosophy and science teaching.* New York: Teachers College Press.

Matthews, M. R. (1994). *Science teaching: The role of history and philosophy of science.* New York: Routledge.

Mayer, R. E. (1987). *Educational psychology.* Dubuque, IA: Little, Brown.

Mertler, C. A. (2006). *Action research: Teachers as researchers in the classroom.* Thousand Oaks, CA: Sage.

Millar, R. (1989). Constructive criticisms. *International Journal of Science Education, 1,* 587–596.

Mills, G. E. (2007). *Action research: A guide for the teacher researcher.* Upper Saddle River, NJ: Pearson.

National Assessment of Educational Progress. (1978). *The third assessment of science (1976–1977).* Denver, CO: Author.

National Council on Measurement in Education. (1995). *Code for professional responsibilities in educational measurement (CPR).* Washington, DC: Author.

National Research Council. (1996). *National science education standards.* Washington, DC: National Academy Press.

National Research Council. (1999). *How people learn: Brain, mind, experience, and school.* Washington, DC: National Academy Press.

National Research Council. (2001a). *Classroom assessment and the national science education standards.* Washington, DC: National Academy Press.

National Research Council. (2001b). *Knowing what students know: The science and design of educational assessment.* Washington, DC: National Academy Press.

National Research Council. (2005). *How students learn: Science in the classroom.* Washington, DC: National Academy Press.

National Science Teachers Association. (1982). *An NSTA position statement.* Washington, DC: Author.

Nersessian, N. (1989). Conceptual changes in science and in science education. *Synthese, 80*(2), 163–183.

Nitko, A. J. (1996). *Educational assessment of students.* Englewood Cliffs, NJ: Prentice Hall.

Nitko, A. J., & Brookhart, S. M. (2007). *Educational assessment of students* (5th ed.). Upper Saddle River, NJ: Pearson.

Novak, J. D. (1998). *Learning, creating, and using knowledge: Concept map as facilitative tools in schools and corporations.* Mahwah, NJ: Lawrence Erlbaum.

Novak, J. D., & Gowin, D. B. (1984). *Learning how to learn.* New York: Cambridge University Press.

Page, E. (1958). *Teacher comments and student performance.* Englewood Cliffs, NJ: Prentice Hall.

Penick, J. E. (1996). Creativity and the value of questions in STS. In R. E. Yager (Ed.), *Science/technology/society as reform in science education* (pp. 84–94). Albany: State University of New York Press.

Pierce, L. V., & O'Malley, J. M. (1992). *Performance and portfolio assessment for language minority students.* Washington, DC: National Clearinghouse for Bilingual Education.

Popham, W. J. (2008). *Transformative assessment.* Alexandria, VA: Association for Supervision and Curriculum Development.

Raizen, S. A., & Kaser, J. S. (1989). Assessing science learning in the elementary school: Why, what, and how? *Phi Delta Kappan, 70,* 718–722.

Richardson, W. (2009). *Blogs, wikis, podcasts* (2nd ed.). Thousand Oaks, CA: Corwin.

Shavelson, R., Baxter, G., & Pine, J. (1992). Performance assessment: Political rhetoric and measurement reality. *Educational Researcher, 21*(4), 22–27.

Shepardson, D. P., & Britsch, S. J. (1997). Children's science journals: Tools for teaching, learning, and assessing. *Science and Children, 34*(5), 13–17, 46–47.

Shepardson, D. P., & Britsch, S. J. (2000). Analyzing children's science journals. *Science and Children, 38*(3), 29–33.

Sloane, K., Wilson, M., & Samson, S. (1996). *Designing an embedded assessment system: From principles to practice.* Retrieved July 22, 2008, from http://bearcenter.berkeley .edu/publications/design96.pdf.

Stiggins, R. J. (1994). *Student-centered classroom assessment.* New York: Macmillan.

Stiggins, R. J. (2008). *Student-involved assessment FOR learning.* Upper Saddle River, NJ: Pearson Education.

Strauss, S., & Stavey, R. (1983). *Educational developmental psychology and curriculum development: The case of heat and temperature.* Ithaca, NY: Cornell University, International Seminar on Misconceptions in Science and Mathematics.

Swindoll, C. R. (1994). *Killing goats, pulling thorns.* Grand Rapids, MI: Zondervan.

Tamir, P., & Amir, R. (1981). High school as viewed by college students in Israel. *Studies in Educational Evaluation, 7*(2), 211–225.

Taylor, P. C., Fraser, B. J., & White, L. R. (1994, March). *A classroom environment questionnaire for science educators interested in the constructivist reform of school science.* Paper presented at the annual meeting of the National Association for Research in Science Teaching, Anaheim, CA.

Thagard, P. (1992). *Conceptual revolutions.* Princeton, NJ: Princeton University Press.

Thomas, J. A., Pedersen, J. E., & Finson, K. (2001). Validating the draw-a-science-teacher-test checklist (DASTT-C): Exploring mental models. *Journal of Science Teacher Education, 12*(4), 295–310.

Torrance, E. P. (1969). *Creativity.* Belmont, CA: Dimensions.

Urban, K. K. (2004). Assessing creativity: The test for creative thinking-drawing production (TCT-DP) the concept, application, evaluation, and international studies [Electronic version]. *Psychology Science, 46*(3), 387–397.

Varrella, G., Kellerman, L., & Penick, J. (1993). In R. E. Yager (Ed.), *Student teaching handbook.* Iowa City: University of Iowa, Science Education Center.

Wheatley, G. H. (1991). Constructivist perspectives on science and mathematics learning. *Science Education, 75*(1), 9–21.

Wiggins, G. (1989a). A true test: Toward more authentic and equitable assessment. *Phi Delta Kappan, 70,* 703–713.

Wiggins, G. (1989b). Teaching to the (authentic) test. *Educational Leadership, 46*(7), 41–47.

Wiggins, G. (1993). *Assessing student performance.* San Francisco: Jossey-Bass.

Wiggins, G. (1998). *Educative assessment: Designing assessments to inform and improve student performance.* San Francisco: Jossey-Bass.

Wiggins, G., & McTighe, J. (2005). *Understanding by design* (2nd ed.). Alexandria, VA: Association for Supervision and Curriculum Development.

Yager, R. E. (1987). Assess all five domains of science. *The Science Teacher, 54*(7), 33–37.

Yager, R. E., & McCormack, A. J. (1989). Assessing teaching/learning successes on multiple domains of science and science education. *Science Education, 73*(1), 45–58.

Yager, R. E., & Roy, R. (1993). STS: Most pervasive and most radical of reform approaches to "Science" education. In R. E. Yager (Ed.), *What research says to the science teacher: The science, technology, society movement* (pp. 7–13). Arlington, VA: National Science Teachers Association.

Index

CORWIN

A SAGE Company

The Corwin logo—a raven striding across an open book—represents the union of courage and learning. Corwin is committed to improving education for all learners by publishing books and other professional development resources for those serving the field of PreK–12 education. By providing practical, hands-on materials, Corwin continues to carry out the promise of its motto: **"Helping Educators Do Their Work Better."**